◊

Garbage Management in Japan:

Leading the Way

GARBAGE MANAGEMENT IN JAPAN:

LEADING THE WAY

by

ALLEN HERSHKOWITZ, PH.D.

and

EUGENE SALERNI, PH.D.

PREFACE BY MAURICE HINCHEY

INFORM, Inc.
381 Park Avenue South
New York, NY 10016
(212) 689-4040

Library of Congress Cataloging-in-Publication Data

Hershkowitz, Allen.
    Garbage management in Japan.

    Bibliography: p.
    1. Refuse and refuse disposal--Japan. 2. Recycling
(Waste, etc.)--Japan. I. Salerni, Eugene. II. Title.
TD789.J3H47   1987   363.7'28'0952   87-82166
ISBN 0-918780-43-8 (pbk.)

INFORM, Inc., founded in 1973, is a nonprofit research organization that identifies and reports on practical actions for the protection and conservation of natural resources and public health. INFORM's research is published in books, abstracts, newsletters and articles. Its work is supported by contributions from individuals and corporations and by grants from over 40 foundations.

Cover design by Saul Lambert
Book design and typography by James Carr and Perrin Stryker
Photographs by Allen Hershkowitz and Eugene Salerni
Map by James Carr
Printed on recycled paper

Illustrations and charts from: *Recycling '86, Recycle Life,* and *Risachan's Picture Diary on Recycling,* by Clean Japan Center ; *Plastic Waste,* by Plastic Waste Management Institute; and *Solid Waste Management and Night Soil Treatment,* by Japan International Cooperation Agency.

TABLE OF CONTENTS

CHARTS

TABLES

PREFACE

Imagine a city where all residents routinely separate their household waste into seven categories. Virtually all newspapers, magazines, glass bottles and metal cans are sorted and recycled. After residents further remove incombustible materials and hazardous wastes, including batteries, the remaining combustible wastes and large bulky items are sent to a "Recycling Cultural Center." There, bulky wastes are screened and salvageable items, such as bicycles and furniture, are extracted and sent to a workshop where they are refurbished by handicapped persons and sold to the public.

The remaining bulky waste is shredded and metals are recovered. Some combustible waste is composted and sent to an adjacent greenhouse where it is used to grow plants for sale to the public. The greenhouse is also staffed by the handicapped, and uses steam from a waste incinerator for warmth when needed.

The incinerator is equipped with acid gas controls and is operated by trained, certified engineers. Its emissions are fully monitored. Its fly ash is cemented to prevent leaching, and mixed with bottom ash for burial in a lined landfill, along with incombustible waste. As it is filled, the landfill will be converted to accomodate recreational facilities. The cultural center also has a community swimming pool (heated by the incinerator) and classrooms where school children and community residents can learn about proper waste management.

In short, this city represents the optimum solid waste management program, at least as conceived by planners in the U.S. Over half of all discards are recycled, with most of the remainder burned for energy recovery. Landfilling is reserved only for treated residues and inert wastes.

There's only one catch.
The city is in Japan.

The evidence presented in the chapters that follow implies, and rightly so, that in waste production and management we are as different from the Japanese as we are in virtually every other aspect of our national economic and cultural life.

On a per capita basis, we generate more than twice the waste of our Japanese friends.

We are six times as dependent on landfills.

We do only a small fraction of the recycling done in Japan. And some of our own recycling actually is done by the Japanese, who take our scrap metal exported from ports like New York and Los Angeles and make it back into pick-up trucks and VCRs.

In Japan, we see a conscious and well-organized effort to minimize waste and to maximize recycling. Virtually every citizen and industry participates in some way. Because discarded materials are not necessarily waste (they may be resources), the utilization of them can help the nation to be less dependent on others for raw materials. The curse of Japanese geography becomes the blessing of its waste management program.

The attention the Japanese pay to their waste sets the standard for the industrialized world. The lessons we learn from examining this standard include these:

That waste-to-energy incineration is not incompatible with high rates of recycling, a fiction promoted by some community groups in the U.S.; in fact, the two assist each other in meeting the higher waste management goal of reducing landfilling.

That local, preferably municipal government control of basic waste management strategies is critical to maintaining public interest, support and participation. Rational waste management, as practiced in Japan, is not merely a technical development, it is a socio-political achievement, and one which requires public accountability by those involved in operating local programs. Too often, in the U.S., local elected officials have abdicated this responsibility and turned their powers over to others whose chief obligation is to their shareholders, whose own wastebaskets may be far, far away.

That waste management need not be set apart from the life of the community. Educating the public is not something to be avoided but something to be sought. If we remember any phrase from this report, it should be the goal of the "eradication of sanitary illiteracy." Such a goal is fundamentally intertwined with one's sense of community, of how we live amongst our fellow citizens. Sanitary illiteracy is incompatible with community well-being.

It is often said, "geography is destiny," and for an island nation with so little habitable land, the impact of geography on waste management strategy is almost total.

But it does not follow that because the United States has vast amounts of land that we do not have a waste crisis, or that we should choose to use our land resources to bury our garbage. If we Americans reject that choice, as I think we do, then we must adopt a strategy of minimizing our waste, maximizing our recycling, reducing volume and toxicity through well-controlled waste-to-energy facilities, and landfilling mainly treated residues and inert waste.

Garbage Management in Japan: Leading the Way, demonstrates that the solutions to the solid waste management problem are not technical. There is no philosophers' stone in our future capable of turning garbage into sea water. The solutions to the waste problem are fundamentally political and cultural.

In the U.S. we are expecting waste-to-energy incinerators to replace our depleted and decrepit landfills, but without widespread public support. Is this because the public rightly believes we have done too little to reduce and recycle? Or is it because the public lacks confidence that our waste-to-energy programs will be compatible with other community goals, clean air, good public health? States in the U.S. have been too slow in requiring certification and training for our operators, a step which could help build public confidence. Perhaps it is because enough has not been asked of our own citizens in helping to address the waste crisis.

What we do with our garbage in the year 2000 will be a matter of our own collective choosing. Some governmental officials are increasingly cognizant of the threat to our economy and environment posed by failing to address the problem, yet political leaders too much fear that the public will not accept the solutions, and so decisions are delayed until after the next election...at least.

We can choose, or we can choose not to choose. That is the choice.

If we want a country which conserves its natural resources, protects the environment and encourages public accountability, then we can go down the path found by the Japanese years ago.

Garbage Management in Japan: Leading the Way makes an invaluable contribution to helping all of us find that path toward responsible waste management and resource conservation. It is both a guide-post and a road map. But in the final analysis, it is we who must decide to make the journey to a city we not only imagine, but can choose to create.

Assemblymember Maurice D. Hinchey
Chairman
New York State Legislative Commission
on Solid Waste Management

ACKNOWLEDGEMENTS

Detailed and accurate research into any nation's solid waste management practices must involve the cooperation of many specialists in government and industry. This was certainly true for two Americans travelling around Japan for five weeks to gather obscure data, conduct sometimes translated interviews and tour industrial facilities to write a report about that nation's unique solid waste management practices. This report is a tribute to the friendly and cooperative response we received from our Japanese hosts and colleagues and from many other people who also deserve our thanks.

We benefited from extraordinary cooperation by Kazunobu Onogawa, Takashi Hayase, Taka Hiraishi, Akira Takimura, and Masahiro Ohe of Japan's Environment Agency and by Saburo Kato of the Ministry of Health and Welfare. Our thanks to all of them for providing us with many lengthy interviews (often late into the night), government reports, office amenities and factual reviews.

Our gratitude also to Kunihiro Nakazato of Takuma Co., Ltd. for assisting us in our travels and for providing us with much valuable information not only on solid waste management, but also on the Japanese way of life. Certain other members of Japanese industry deserve our thanks for the information they provided and the access they provided to their facilities: Kenzo Shiomi, Mamoru Yamada, Koichi Kondo, Norio Hiroshima and Tomotsu Matsumara of Kawasaki Heavy Industries; Mark Matsuoka, and R. Kenneth Merkey of C. Itoh & Co.; Yasujiro Wakamura of Itoh Takuma Resources Systems, Inc.

At the Clean Japan Center, Hidenobu Ogasawara provided us with volumes of data and useful information, and Dr. Masaru

xiv   GARBAGE MANAGEMENT IN JAPAN

Tanaka of the Institute of Public Health was also helpful, as was Dr. Masakatsu Hiraoka of Kyoto University.

Special thanks are owed to Hiroyuki Hatano and Mari Hatano of Kyoto University for inviting us to participate in the highly successful and illuminating Machida Combustion Workshop. At Machida we were aided greatly by the information provided to us by Kiyoshi Saeki, Manager of the Recycling Cultural Center, as well as by Drs. David Hay and Ray Clement of Environment Canada, Dr. Otto Hutzinger of Bayreuth University, Dr. Warren Crummett of Dow Chemical Company and, above all, by Dr. Frank Karasek who arranged with Dr. Hiroyuki Hatano that we be invited to participate in the Combustion Workshop. Thanks also to Junko and Noriko.

Particular thanks are owed to members of Japan's Plastic Waste Management Institute, including Hidehiko Mononaga, Shiro Hamaya, Masao Masuno, and Makoto Hiraoka. And to the more than 25 skilled and well-trained operators we spoke with at the eight waste-to-energy plants we visited, we offer our respect and gratitude.

While in Japan we also attended the Sixth International Dioxin Conference where we benefited from numerous and lengthy conversations with Dr. Christoffer Rappe of Sweden's Umea University, Drs. Olle Aslander and Bo Jansson of the Swedish Environmental Protection Board, David Sussman of Ogden Martin Systems, Dr. Ted Goldfarb of Stony Brook University and Dr. Paul Connett of St. Lawrence University. Thanks also to Keiko.

To the staff at INFORM, we owe many thanks. A very great debt is owed to Nancy Lilienthal, who meticulously reviewed this document on a number of occasions and, in cooperation with Perrin Stryker, made this a much more logical, accurate and easy report to read. Joanna Underwood, as always, provided unbounded enthusiasm as well as substantive review. We owe especial thanks to Ellen Poteet, who helped research and arrange our itinerary, with the able assistance of Ellen Miller. We benefited importantly as well from the hard work of Maarten de Kadt, Dawn Begian, James Carr, Charles Lowy, John Mensing, Mark Brown, Irmgard Hunt, Suzanne Wilson and Roger Miller.

Thanks also to Peter Green for his fine editing of the manuscript, to Mari Ihara who provided expert translation of some obscure and technical reports, and to the Fund for the City of New York and its Nonprofit Computer Exchange for its help and cooperation in the report's production.

The authors also benefited from the encouragement and active support of Gordon Boyd and Stephen Wilson of the New York State Legislative Commission on Solid Waste Management. And a number of our American colleagues also contributed to the quality of our research agenda and this report, before we left or after we returned, either in conversation or by reviewing drafts of the manuscript. These individuals include Neil Seldman, Eric Goldstein, Barry Mannis, Floyd Hasselriis, Charles Johnson, Lewis Cohen, and Marjorie Clark. Thanks also to Meg, Dylan, Ellen, Stephen, and Karen, who made it all a lot easier.

Finally, the authors would like to thank the North Shore Unitarian Universalist Veatch Program, which made a special contribution to INFORM's Japan research. This contribution, along with the generous contributions of other donors to INFORM's Solid Waste Management Research Program, have given us the total support needed to advance our research in this critical field. These donors include: Robert Sterling Clark Foundation, The New York Community Trust, Marilyn M. Simpson Charitable Trust, Mary Reynolds Babcock Foundation, Geraldine R. Dodge Foundation, Victoria Foundation, The Fund for New Jersey, and German Marshall Fund of the United States.

IMPORTANT NOTES FOR THE READER

The definition of municipal waste.

That used by the Japanese government and industry differs fundamentally from the definition used in the U.S. and Europe. This affects the understanding of data presented in some charts and tables.

The Japanese call municipal solid waste the material that, after recycling, requires treatment and disposal by the municipality. This does not include industrial wastes. In Japan, materials that are recycled are considered as resources, not wastes, and are not listed in solid waste data.

Not taking account of this difference in definition would result in overstating by as much as 100% the amount of waste incinerated and landfilled in Japan. When Japanese data indicate that 68% of the nation's waste is incinerated, 30% is directly landfilled and 2% is disposed of through other means, the reader should note that these figures are for wastes that remain after recycling. In western terms, Japan recycles about 50% of its waste, incinerates about 34%, and directly landfills (or otherwise disposes of) 16% In this report all aggregate solid waste data do not include recycled material unless otherwise stated.

Materials Recovery and Recycling.

Chapter Four indicates that not all the materials separated from Japan's municipal waste stream are recycled. Materials that are not going to the incinerator are separated for landfill or another form of disposal. These materials include non-combustibles, some glass, tires, batteries and some plastics.

The process of sorting non-recyclable, non-combustible materials from the waste stream to prevent them from going to an incinerator is referred to in this text as "materials recovery." The process of using recovered items in the manufacture of new products is, of course, referred to as recycling.

Tonnage figures.

Those used in Japan and in this report are in metric units. A metric tonne (t) is 1,000 kilograms and is equivalent to 2,204 lb. or 1.1 U.S. tons. Conversely, one U.S. ton is equivalent to 0.9 metric tonnes. All tonne per day (t/d) figures are based on 365 days per year (See Conversion Chart, Appendix 1).

Japanese currency.

The basic unit of Japanese currency is the yen (pronounced "en"). All conversions used in this report are based on the exchange rate, during INFORM's fall 1986 visit, of 154 yen per $1. The reader is advised to use caution when interpreting cost data since the value of the yen fluctuates. At the time of publication, in the fall of 1987, the yen is particularly strong. As recently as 1985, the rate was 240 yen to the dollar.

Japanese annual waste statistics.

Unless otherwise indicated, these are based on a fiscal year (FY) that begins April 1 and ends March 31.

Japan's Ten Most Populated Prefectures

Japan's population density has intensified it's garbage management problems. About 60% percent of Japan's 120 million citizens live in areas with population densities greater than 4,000 per square kilometer and 45% of the entire Japanese population lives in the three metropolitan areas of Tokyo, Osaka and Nagoya (Aichi prefecture), which together comprise only 3% of the land.

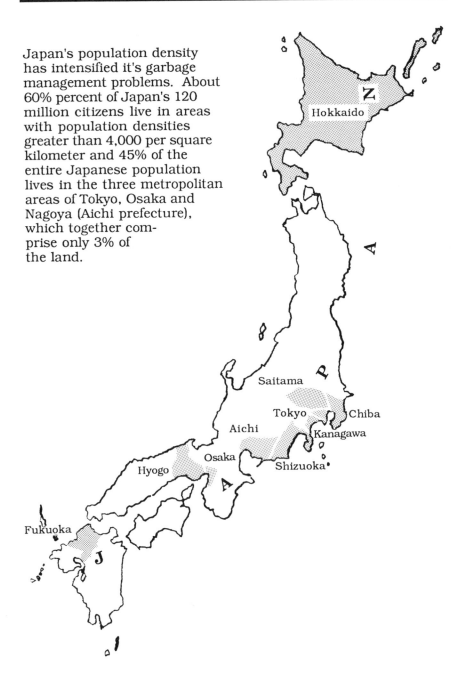

I

INTRODUCTION

The environmental concerns related to solid waste management in the U.S. have led many people and government agencies to look abroad for information about recycling, garbage burning and, more generally, the design of waste management programs.

INFORM visited four European countries to study garbage burning and in 1986 published *Garbage Burning: Lessons from Europe.* Following this, INFORM looked to see if U.S. communities could also benefit from information about waste management in Japan. We discovered, based on discussions with members of Japanese industry and government, that America could indeed learn from the Japanese much about the technology and design of integrated solid waste management.

SOLID WASTE MANAGEMENT PROBLEMS IN THE U.S.

The U.S. has reached a watershed in its handling of municipal wastes. The traditionally used, yet unhealthful, practices of indiscriminately dumping unprocessed municipal waste into landfills is now taking its toll: in groundwater and surface water pollution, gas emissions, enormous clean-up costs and a socially divisive search for alternative management, treatment and disposal policies.

In as many as 40 states, landfills are being ordered to close because they have reached capacity or because they threaten to pollute groundwater. And new landfills are increasingly difficult to site, due to "not in my backyard" sentiment. About 200 of New York's 500 landfills closed between 1983 and 1987. New York City and the City of Los Angeles each have only one remaining major landfill, and soon all of Seattle's landfills will be closed.

1

Hundreds of similar situations throughout the U.S. could be found by reading the front pages of the nation's newspapers.

Landfills that accept unprocessed municipal wastes can pollute groundwater and surface water with the toxic materials found in ordinary waste, including metals such as lead and mercury. And once polluted, the water supplies tainted by landfills are extremely difficult to clean up, if they can be cleaned at all. Moreover, while the timetable for cleaning up the nation's closed landfills remains uncertain, problems exist even where landfills remain open. Of 9,284 landfills in the U.S. that accept municipal solid wastes, only 59% have permits or approved plans for operating. And even those landfills with permits are behind in the technology of environmental protection. Only 15% have liners to channel the toxic leachate that landfills produce; and even fewer channel, collect and chemically treat these poisons.

While federal and state regulatory agencies have been ordering the closure of landfills, they have been slow in identifying solid waste management directions that satisfy the basic environmental concerns of communities that lose their landfills. One response to the closing of landfills is to burn more garbage. About 75 high-volume garbage burning plants are now operating around the U.S., and over 140 more are under construction or planned.

However, this complicated and controversial garbage burning process generates substantial environmental concerns. Burning produces emissions of acid gases such as hydrochloric acid, heavy metals such as mercury, lead and arsenic, and toxic organic compounds such as dioxins and furans. And burning does not eliminate the need for and concern about landfills since it generates potentially hazardous ash that must be dumped.

Seeking to avoid the environmental costs associated with landfills and burning, hundreds of communities are looking to recycle their waste. An estimated 8%-10% of all U.S. municipal waste is now recycled and perhaps as many as 8,000 recycling programs are now operated in the U.S. At least three states, (Oregon, Rhode Island and New Jersey), have made some recycling mandatory. Unfortunately, as the data in this study indicate, communities cannot recycle their entire waste stream, even under the most encouraging circumstances.

The industrial processes for recycling materials also generate pollutants that raise environmental concerns. According to a

Swedish chemist, Dr. Christoffer Rappe, perhaps as much as 25% of all background levels of dioxin are associated with industrial processes that include waste recycling, such as paper pulping and bleaching, copper refining and steel production. Alas, all municipal solid waste management strategies raise some environmental concerns, and communities are in desperate need of reliable information to minimize the adverse environmental effects.

THE JAPANESE APPROACH

Japan has a well-developed municipal solid waste management program that benefits from a great deal of social cohesion and a high degree of active coordination among the national, regional and local governments. Unlike the U.S., where solid waste management is almost exclusively the domain of state and municipal governments, in Japan the national government plays an active role in shaping municipal solid waste management through subsidies and regulations. The national government in Japan also mandates and helps to coordinate the collection of waste data needed to most efficiently manage municipal wastes. No comparable role is played by the U.S. government. →

Japan's solid waste management program is partly imposed by geography. The nation has rapidly become urbanized, with a population one-half that of the United States living in a land area equal to California, but with only 21% of that land habitable. About 60% percent of Japan's 120 million citizens live in areas with population densities greater than 4,000 per square kilometer and 45% of the entire Japanese population lives in the three metropolitan areas of Tokyo, Osaka and Nagoya, which together comprise only 3% of the land. As a result, the availability of land for the continued disposal of waste is at a premium and the nation has only a few years of total landfill capacity left.

Moreover, Japan is heavily dependent on imports and, consequently, the value of refuse as a resource is widely recognized. These factors combine to shape Japan's waste management policies and result in the implementation of many creative and sophisticated techniques. Japan now experiences extraordinarily high levels of materials recycling, over 95% in the case of some commodities, and has built over 1,900 incinerators during the past 25 years to process two-thirds of the remaining municipal waste.

TEN JAPANESE PROBLEMS

One reason Japan is so far advanced is that early on it had to face problems that the U.S. is now facing. These are (as identified by the Environment Agency Minister's Advisory Council):

1. A growing difficulty in siting waste disposal facilities owing largely to scarcity of available land

2. Concern about dioxin emissions from all incinerators, and a particular concern about air pollution from older (and smaller) incinerators without emissions control technology

3. Difficulties in gaining understanding and cooperation of residents for appropriate waste management

4. An increase in waste disposal problems owing to the increase in plastics, batteries that contain mercury, and bulky household appliances

5. Littering of soft-drink containers

6. An increase in waste management costs due to increased collection costs and the introduction of sophisticated waste treatment technology

7. Illegal dumping

8. The need for still more recycling

9. Wastewater pollution from incinerators

10. Unreliable procedures for constructing reliable environmental impact statements.

Because all these are problems also being faced in the U.S., the methods that the Japanese are using to resolve these issues should be useful to U.S. planners, despite the cultural differences between the U.S. and Japan.

◊

II

FINDINGS

WASTE GENERATION AND DISPOSAL

➤ The Japanese maintain extraordinarily precise data on municipal solid waste generation, recycling, materials recovery, incineration and landfilling by locality, region, prefecture and nationally. This resource accounting system is supplemented by materials recovery data supplied by private industry, and works to help the Japanese efficiently allocate capital, labor and land to manage solid wastes. (Chapter Three)

➤ Wastes are disposed of in three major ways after recycling, which may reprocess as much as 50% of Japan's wastes. Incineration treats 68% of the remaining wastes while landfills accept 30%. Another 2% are handled by "other" means, including composting. (Chapter Three)

➤ After recycling, only half as much waste per person is generated in Japan as in the U.S. The Japanese generate 2.2 lb. of garbage per person per day with commercial wastes included, and 1.8 lb. per person per day of residential waste. Even so, a national campaign to encourage cities to reduce waste through recycling is under way. (Chapter Three)

RECYCLING

➤ Recycling is Japan's "most desirable" waste management method. There are five reasons for the importance and the extraordinary success of recycling: 1) a long history of recycling; 2) the country's enormous reliance on imported primary raw materials; 3) the need to control pollutants from landfills and incinerators; 4) government support; and 5) a wide-ranging public education program. (Chapter Five)

➤ The Japanese believe that social and economic factors are more important to the success of recycling than whether or not an item is technically recyclable. These factors place a limit on how much recycling can be accomplished. (Chapter Five)

➤ Most collection of recyclables is carried out by volunteer civic groups and private resource recovery dealers. Collection programs operated by municipalities account for a smaller percentage of recycling activities. (Chapter Four)

➤ The use of composting is declining, and less than 0.2% of municipal solid waste is now composted. (Chapter Five)

➤ The Japanese describe plastics as a "waste difficult to be disposed of" and about one-half of Japan's municipalities have special handling procedures for this material, such as recycling or landfilling it to avoid incineration. (Chapter Five)

➤ Over 2,300 of Japan's 3,255 municipalities pre-sort certain non-automotive batteries prior to landfilling or incineration out of a concern — not substantiated by any incinerator-specific health-effect data — about the pollution hazards they might generate. Forty-seven of these municipalities go so far as to solidify the batteries in cement and then landfill them. (Chapter Five)

➤ The country recycles about 50% of its paper. For each ton recycled, the Japanese estimate that 20 trees are saved. In 1983 Japan recovered more than 9.1 million tons of paper, saving over 180 million trees. (Chapter Five)

➤ About 42% of glass bottles are made from recycled cullet, and 66% of all bottles are reused an average of three times. About 95% of some bottles, such as beer bottles and 2-liter sake bottles, are reused an average of 20 times. (Chapter Five)

➤ Over 40% of all steel and aluminum cans are recycled, providing significant annual energy savings. The associated cost savings help make Japan's exports of these metals more competitive. (Chapter Five)

INCINERATION

▷ Over 1,900 incinerators have been built in the past 25 years despite Japan's well-organized and successful recycling programs. /And 361 incinerators, representing 69% of the incineration capacity in Japan, recover energy. Apartment house incineration of wastes is not common. (Chapter Six)

▷ All incinerator plants are publicly owned, although one-third are operated by private contractors. (Chapter Six)

▷ Waste incineration is primarily designed and operated as a waste disposal process and is only secondarily a means for energy recovery./ Japan's regulators say that unless such a position is maintained in their worker training programs, an inadequate sensitivity to the hazards of handling and burning garbage might occur. (Chapter Six)

▷ According to Japanese industry and government representatives, opposition to garbage burning in Japan exists "everywhere" because of "many failures in the past," a concern about property values, and truck traffic. This is alleviated by the local and national government by upgrading regulations and plant equipment and practices and by providing amenities to the community in exchange for siting. (Chapter Six)

▷ At four of the eight incineration plants visited by INFORM, ash was cemented or remelted to reduce the likelihood of heavy metals leaching after the ash is deposited in a landfill. Government documents suggest this to be a widespread practice. Japan's Environment Agency states that landfills containing briquetted ash can be more easily reclaimed than landfills that accept unprocessed ash. The national regulation for ignition loss is below 5% (by volume) of the material entering the furnace, for plants of 200 t/d or greater. (Chapter Six)

▷ Many plants in Japan have telemetering systems for direct communication between environmental regulators and the plant. Under extremely adverse ambient air conditions, plants can be requested to temporarily suspend operations. (Chapter Six)

▷ The government believes no health threat exists from dioxin, but Japanese health effect calculations and dioxin emissions data for incinerators are subject to international dispute. (Chapter Six)

➤ At least five national laws require specialized training of workers at incineration facilities. Workers at recycling centers and landfills also receive specialized training. Despite this, the accident rate in Japan's sanitation industry, though decreasing, is greater than in any other industrial sector. (Chapter Six)

➤ National incinerator air emission regulations are less comprehensive and strict than those in some Western countries, but regional and municipal regulations are often more demanding. Control of sulfur dioxide (SO_2) and oxides of nitrogen (NO_x) are based on regional ambient air quality assessments, as well as on how well the control technology is known to perform. Prefectural governors work together to reduce regional pollution problems. (Chapter Six)

➤ Regulatory standards tend to be based on maximum prevention of pollution even when health effects data are uncertain or unavailable. In practice, regulations tend to be based on the government's assessment of how well technology can perform and the environmental concerns of the local population. They ignore the difficult-to-obtain correlations between health effect data and emissions from a specific plant. (Chapter Six)

➤ The Japanese government subsidizes 25% of the initial capital costs for constructing municipal solid waste incinerators and uses the subsidy as an incentive to achieve maximum pollution control. In areas where pollution is especially severe, the government subsidy can be as high as 50%. (Chapter Six)

➤ The government views wastewater control at incineration plants as essential, especially as wet scrubbers are increasingly used to control acid gases. All waste incinerators in Japan with a capacity of five tonnes/day or more are subject to the Water Pollution Control Law. (Chapter Six)

➤ Emissions tests for at least four common pollutants are required at municipal waste incinerators at least once every two months. (Chapter Six)

LANDFILLS

> Landfill capacity is disappearing, and in 1983 Japan had only six to seven years of capacity left in existing landfills. (Chapter Seven)

> Landfills are designed to prevent groundwater contamination using three measures that can most effectively prevent pollutants from leaching into groundwater: impermeable liners, leachate collection and wastewater treatment. (Chapter Seven)

> Only 10%-20% of Japan's household garbage goes to landfills unprocessed. (Chapter Seven)

III

JAPAN'S SOLID WASTE MANAGEMENT

LAND USE PRESSURES

Japan was probably the first major industrialized state to confront the problem of decreasing landfill space. A document on solid waste management issued in 1986 by the Japan International Cooperation Agency (JICA) says the need to reduce the volume of garbage going to landfills emerged immediately "after World War II, [when] waste increase[s] and urbanization made [it] ... difficult to rely on land disposal ... [for] waste management."

Between 1950 and 1985, Japan's population increased from 83 million to 120 million, with over 45% of the population concentrated in three major urban areas, Tokyo, Osaka and Nagoya. The extreme concentration in these three areas that comprise only 3% of Japan's total land, is largely because only about 21% of the land is habitable (80,000 out of a total 378,000 square kilometers (km^2). The rest is mountainous and forested. In the habitable area, Japan's population density is among the most concentrated in the world, with 1,452 people per km^2 (see Table 1). In contrast, population density in the habitable area of the U.S. is 50 people per km^2, in France it is 158, in Britain 358 and in W. Germany 386 people per km^2 (See Table 1).

The Clean Japan Center, a private foundation receiving both government and industry support, stated in 1986 that as a result of Japan's population density "the availability of [landfill] sites for final disposal [of waste] is insufficient all over the country, and is becoming more difficult to secure." According to Mr. K. Ono of Japan's Kyosanto (Communist) party, "in the urban areas garbage is one of the most serious problems."

Consequently, because of the environmental problems caused by landfilling unprocessed household garbage, most non-recycled

11

TABLE 1 Area and Population Density in 1983

(Population per Square Kilometer)

	Japan	U.S.A.	West Germany	France	Britain
Total area (100km²)	3,777	93,718	2,486	5,470	2,471
Habitable area (100km²)	805	45,814	1,594	3,389	1,564
Ratio of habitable to total (%)	21	49	64	62	63
Population density in total area (person/km²)	320	25	248	99	229
Population density in habitable area (person/km²)	1,452	50	386	158	358

Source: The Plastic Waste Management Institute

combustible wastes are incinerated. Dr. M. Tanaka, Head of the Solid Waste Division at the Institute of Public Health, stated that "after recycling, we want to incinerate 100% of the [non-recycled] combustible material...There is no choice."

LAWS GOVERNING WASTES

After World War II, Japan sought to limit the amount of garbage going to its depleting and hard-to-site landfills. In 1954 the Public Cleansing Law was enacted to reduce the volume of garbage and encourage sanitary waste treatment "through [the] construction of solid waste incinerators." Planning was needed because throughout the 1960s and 1970s the generation of residential waste grew substantially, from about 55,000 to almost 100,000 tonnes/day (t/d) (See Chart 1).

In 1983, the average daily per capita residential waste generation rate was 1.8 lb (826 grams) per day or 2.2 lb if commercial waste is included. This is approximately half the U.S. rate, mainly because of Japan's recycling efforts. However, waste generation varies by prefecture, due in part to differences in urban density. And in large cities, where the land available for landfilling waste is most scarce, the per capita generation of residential waste is even greater than the national average, sometimes by as much as 100% (Tables 2 and 3).

CHART 1 Generated Amounts of Residential Municipal Wastes

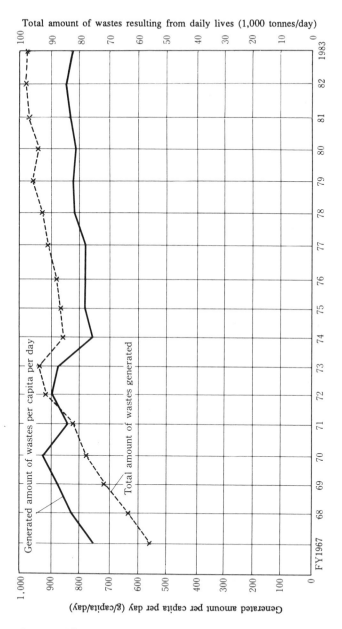

Total amount of wastes resulting from daily lives (1,000 tonnes/day)

Source: Plastic Waste Management Institute (PWMI)

TABLE 2 Daily Residential Waste Generation (by prefecture, 1983)

	Population (thousands)	Discharge per Capita per Day (grams)
Tokyo	11,664	1,383
Osaka	8,563	1,023
Kanagawa	7,226	831
Aichi	6,326	799
Hokkaido	5,682	970
Saitama	5,677	660
Hyogo	5,253	714
Chiba	5,003	727
Fukuoka	4,647	774
Shizuoka	3,541	653
Hiroshima	2,807	712
Ibaraki	2,667	709
Kyoto	2,559	791
Niigata	2,478	776
Miyagi	2,138	722
Nagano	2,108	687
Fukushima	2,075	645
Gifu	2,007	614
Okayama	1,911	663
Gunma	1,902	704
Kagoshima	1,825	690
Kumamoto	1,823	568
Tochigi	1,806	700
Mie	1,725	796
Nagasaki	1,607	759
Yamaguchi	1,603	693
Aomori	1,563	911
Ehime	1,541	699
Iwate	1,457	738
Nara	1,270	747
Yamagata	1,268	553
Akita	1,263	783
Oita	1,241	651
Miyazaki	1,180	743
Okinawa	1,164	713
Ishikawa	1,152	775
Shiga	1,132	679
Toyama	1,119	748
Wakayama	1,102	890
Kagawa	1,013	678
Saga	885	689
Tokushima	842	835
Kochi	835	744
Wamanshi	825	621
Fukui	809	727
Shimame	796	699
Tottori	617	917
TOTAL	119,733	826

Source: Clean Japan Center

TABLE 3 Daily Residential Waste Generation
(some selected cities, 1982)

City	Population (thousands)	Discharge per Capita per Day (grams)
Tokyo	8,334	1,710
Osaka	2,623	1,806
Kyoto	1,473	1,021
Sapporo	1,470	1,773
Fukuoka	1,118	1,300
Kawasaki	1,097	1,288

Source: Solid Waste Management and
Night Soil Treatment, Vol. I - JICA

The Japanese have identified a relationship between socio-economic status and per capita waste generation. The higher generation of wastes in urban areas is explained by the higher economic position of urban residents, more commercial activity and the increased availability of heavily-packaged consumer items. In rural areas there are greater opportunities for self-disposal of wastes. According to JICA,

Generally speaking, the higher the income and the [greater] the family size, the more solid waste and the more variety of solid wastes that are generated.

Japan's increasing waste disposal burden led to the passage in 1976 of the Waste Disposal and Public Cleansing Law, an amendment to Japan's 1954 Public Cleansing Law. Both laws promoted incineration and the turning away from unsafe landfills 1) to prevent the pollution of groundwater and public water areas caused by leachate from landfill sites and 2) to reduce the volume of waste going to landfills and to stabilize the property of waste.

To help implement their waste disposal laws, the Japanese maintain precise data on wastes generated, materials recycled, materials recovered, and wastes incinerated and landfilled by locality, prefecture and nationally. JICA explains:

Estimates of [the] quantity and quality of waste to be disposed [of] are essential for arranging [and sizing] the equipment required for storing, collecting, transporting and disposing [of] waste, preparing materials and reserving workers and areas for disposal sites.

In 1983, the last year for which complete data are available, after recycling, 67.6% of Japan's waste was incinerated; 29.6%

directly landfilled, excluding incinerator ash, and 2.8% composted or disposed of by other means. Incineration of wastes has
been steadily increasing and as much as 88% of Japan's wastes
may be burned by the turn of the century (See Chart 2 and Table
4). In contrast, after recycling, the U.S. landfills about 90% of its
municipal solid wastes and incinerates the rest.

TABLE 4 Residential and Commercial Solid Waste Disposal*

	1979	1980	1981	1982	1983
Total waste disposed of					
(t/d)	115,158	113,728	110,209	115,256	110,976
Incineration					
(t/d)	67,887	68,739	71,102	75,264	75,022
(%)	59.0	60.4	64.5	65.3	67.6
Landfill					
(t/d)	44,509	42,139	35,651	37,261	32,842
(%)	38.7	37.1	32.3	32.3	29.6
Compost & other					
(t/d)	2,762	2,850	3,456	2,731	3,112
(%)	2.3	2.5	3.2	2.4	2.8

* After recycling.
Source: JICA

JURISDICTIONAL AUTHORITY FOR COLLECTION, TRANSPORTATION AND DISPOSAL

Although municipalities throughout Japan have equal powers
and responsibilities for waste management, they do not use iden-
tical waste disposal strategies. In fact, the Japanese believe that
allowing municipalities the greatest authority regarding waste
management generates the most locally appropriate mix of waste
management options. JICA states:

> Obviously, different municipalities utilize different refuse disposal
> systems. The reason: municipalities have inherently different eco-
> nomic conditions, social background, and natural environments.
> [Therefore], they use varied evaluation systems. In short, there is
> no need for municipal entities to adopt similar disposal systems.
> [Overall] the purpose of waste disposal is to remove wastes and
> sewerage from our living environment as quickly as possible and to
> make them harmless and stable.

CHART 2 Trends of Residential and Commercial
Waste Disposal Methods

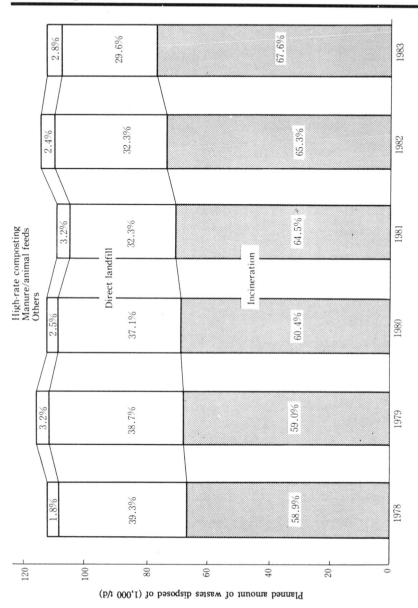

Source: PWMI

Japan is divided into 47 administrative prefectures, each with elected governments and jurisdictional political powers similar to the 50 states in the U.S.A. Within these 47 prefectures are 3,255 municipalities. According to the Basic Law for Environmental Pollution Control, each level of government — municipal, prefectural, and national — has distinct obligations regarding waste disposal, as does the business sector (See Box, below).

Responsibilities of Business Operators and Governments for Municipal Solid Waste Disposal in Japan.

Business operators (commercial and industrial) are required to:

➤ arrange for the disposal of wastes resulting from their business activities;

➤ make efforts for reducing the volume of their wastes and take necessary actions to prevent the products and containers involved in their manufacturing, processing, sales and other business activities from becoming a waste difficult to be disposed of.

Municipalities are required to:

➤ undertake public education regarding safe and clean solid waste disposal;

➤ collect, transport and dispose of domestic waste discharged in their administrative areas according to a Waste Disposal Plan;

➤ carry out domestic waste management efficiently by improving the technological performance of the incineration and landfill facility, consolidating disposal facilities and helping to advance recycling, incineration and landfill techniques;

➤ record and maintain data concerned with domestic waste management;

➤ take measures in line with the national policy for environmental pollution control taking into account the specific [local] natural and social conditions of the areas concerned.

According to Japan's 1976 Waste Disposal Law, the 3,255 municipal governments "have the principal responsibilities for municipal waste management," including authority over operations and environmental controls at incinerators and landfills, as well as record keeping and the collection and transporting of wastes to disposal facilities. In addition, a number of "intermunicipal organizations" have been formed by several municipalities to jointly carry out some aspects of waste management.

Prefectures are required to:

> provide municipalities with necessary technological assistance and take necessary measures for adequate waste disposal by monitoring the progressing status of waste management practices.

The National Government is required to:

> control the emission of pollutants responsible for air, water and soil pollution, establish standards and take measures to deal with noise, vibration and offensive odor;

> establish systems for surveillance, monitoring, measurement, examination and inspection in order to evaluate environmental pollution and to ensure adequate enforcement of pollution control measures;

> promote development of waste management technologies;

> carry out surveys and investigations concerning environmental protection control so as to give necessary technical and financial assistance to municipalities and prefectures.*

* (Currently the national government subsidizes 25% of the expenses for incinerator and landfill construction and most of the rest is covered by municipal loans guaranteed by the national government. In areas of severe pollution, government subsidies can go as high as 50%. See Regulations section of Incineration chapter)

Source: Japan International Cooperation Agency, and the Plastic Waste Management Institute of Japan

Telemetering systems connect a garbage-burning plant to the local environmental agency to help assure compliance with regulations.

Photo: Takuma Co.

Classrooms in garbage-burning plants are used to teach children about their community's solid waste management practices.

Handicapped groups in Machida use compost from garbage wastes to make botanical arrangements for sale to help support handicapped group programs.

A greenhouse of flowers growing in garbage-derived compost supplied by handicapped groups in Machida.

Eradicating Sanitary Illiteracy

The Japanese International Cooperation Agency specified in 1986 at least six Basic Concepts that facilitate widespread acceptance of and participation in a highly coordinated solid waste program. These concepts illustrate the unique social mechanisms for managing solid waste and highlight Japan's emphasis on using public education, community participation, housewives and female professionals to manage its waste and eradicate sanitary illiteracy. (emphasis in original):

1. [Solid waste management projects should be carried out] as a component of a *comprehensive community development program.* Join the project ... with some other projects [having] higher priorities in order to make the project more attractive for the community and to improve the cost-effectiveness of the project ... [The] coordination of various [other community] projects ... such as housing, transportation, employment, drinking water, electricity, etc. ... is indispensable to raise the cost-effectiveness of solid waste management.

2. Activate *communal participation in the project.*

    ➤   so that the selected system will be acceptable for the community economically and socio-culturally.

    ➤   [widespread] participation in the operation and the maintenance of the [waste collection transport and disposal] system ... [will] lower the operation cost.

    ➤   source separation followed by communal recycling activities will reduce the quantity of refuse [needing] to be collected by [the] municipality.

    ➤   not only the ... residents but also the whole community (local industries, rotary club, lion's club, etc.) should participate in the [solid waste] project.

3. Provide *public education* programs in order to promote communal participation.

    ➤   mass [media] campaign (newspaper, radio, television, etc.)

➤ campaign directed to target population with special objectives.

➤ sanitary education curriculum in compulsory education system.

➤ use of audiovisual media which can infiltrate the message into illiterate population. Eradication of sanitary illiteracy is necessary....

4. Respect the *opinions of housewives* ... and utilize *female professionals* as much as possible in the [solid waste] project.

➤ Role of housewives (generally they manage refuse at [the] household level and take care of family members [with] respect to their health).

➤ Role of female professionals (they are more sensitive to the problems of housewives and they will enjoy more community acceptance in field surveys).

5. Select *appropriate technology.*

➤ selected waste collection, transportation, treatment and final disposal systems should be technologically, economically and socio-culturally appropriate and feasible.

6. Arrange *financial resources* for the refuse management system.

➤ steady functioning of the [solid waste] system depends on the existence of firm financial resources. Generally speaking, direct beneficiaries [residents] should finance, at least, the system's operating cost. [However] indirect beneficiaries [local industries and businesses] may bear the capital cost [of communal collection containers, etc., not facility construction].

➤ to lower solid waste management costs, maximize the reduction of waste ... to be handled by [the] municipal refuse service by employing source separation followed by communal recycling activities.

TABLE 5 Collected Amounts of Wastes by Collector (tonnes per day)

Classification	1978	1979	1980	1981	1982	1983
Municipal organizations	56,357	58,164	55,720	56,317	57,045	55,628
	(65.9%)	(65.6%)	(63.0%)	(60.0%)	(60.7%)	(59.7%)
Licensed private contractors	29,134	30,501	32,674	37,607	36,767	37,580
	(34.1%)	(34.4%)	(37.3%)	(40.1%)	(39.3%)	(40.3%)
Total *	85,491	88,665	88,394	93,924	93,902	93,208

*Does not include self-hauled waste.
Source: Derived from data supplied by PWMI.

Many municipalities issue contracts to private companies for the collection of wastes and recoverable materials. The percentage of waste collected by licensed collectors, or private businesses has been increasing since the late 1970s and reached almost 40% of all municipal wastes handled in 1981 (See Table 5).

IV

MATERIALS SEPARATION
AND COLLECTION

MATERIALS SEPARATION: FOUR MAIN CATEGORIES

Over 90% of all of Japan's municipalities divide waste into at least two general categories, combustibles and non-combustibles to, at a minimum, facilitate incineration (reduce load, abrasion, emissions and ash volume) and make landfills safer and longer lasting. Also, within the category of non-combustibles are at least three sub-categories of items that are regularly sorted, including wastes to be landfilled, hazardous household materials and recyclables. Thus, the Japanese separate municipal wastes into four main categories:

1. Recyclable and reusable materials.

   Includes glass, metals and paper.

2. Hazardous materials.

   May pose serious environmental threats if they are incinerated or landfilled. Includes batteries, and other items containing mercury and/or cadmium. These materials are reprocessed or stored until safe disposal or reuse methods become available.

3. Landfilled wastes.

   Non-recyclable, non-hazardous wastes that are also non-combustible, such as broken ceramic items, construction debris and, often, plastics.

4. Incinerated wastes.

   Non-recyclable, non-hazardous, combustible wastes such as soiled paper, kitchen wastes, filmy plastics of mixed resins and unusable wood.

In the home, depending on the municipality, waste can be sorted into additional sub-categories. At the extreme is Zentsuji City which requires waste to be separated into 21 categories.

WHY MATERIALS ARE SEPARATED

Separating wastes and recyclable materials into four classifications achieves both environmental and economic benefits. Materials are separated for three major reasons:

1. to conserve resources in order to reduce Japan's dependence on imported raw materials;

2. to save landfill space and reduce pollution from landfills;

3. to make incinerators less costly and minimize their pollution.

1. Conserving resources

JICA points out that the "level of [recycling] recovery depends almost entirely on economics." Japan has always been heavily dependent on foreign raw materials to supply its industries. Between 1980 and 1985 it imported 99% of its iron ore, 65% of its timber, 18% of its pulp, 95% of its copper, and 78% of its lead. It also imports 99.8% of its oil, which meets about three-fourths of Japan's overall energy requirements (See Table 6). Recycling wastes such as metals, glass, and paper helps to reduce these substantial and costly imports, including energy, which the Japanese government views as making the nation more vulnerable to international pressures.

While municipalities promote recycling at the local level, the Clean Japan Center promotes recycling at the national level by publishing brochures that are widely distributed to educate the public about Japan's "reliance on overseas imports for a great variety of resources" (see Appendix 2). A booklet published by the Ministry of International Trade and Industry (MITI), one of the sponsors of the Clean Japan Center, states:

Materials cannot be had just for the wishing. Lacking in natural resources, Japan can no longer look forward with equanimity to what the future may hold in store...Japan is hard pressed to find substitutes. It is logical for us to try to make good use of the waste. This produces two results: it cleans up the environment, and at the same time recovers precious resources.

TABLE 6 Rates of Imports for Major Commodities (in %)

Commodity	1980	1981	1982	1983	1984
Natural Rubber	100.0	100.0	100.0	100.0	100.0
Nickel	100.0	100.0	100.0	100.0	100.0
Raw Cotton	100.0	100.0	100.0	100.0	100.0
Wool	100.0	100.0	100.0	100.0	100.0
Crude Oil	99.8	99.8	99.8	99.8	99.8
Iron Ore	99.6	99.5	99.6	99.6	99.7
Copper Ore	94.3	94.8	95.4	95.1	95.3
Coal for coke making	89.9	91.3	92.4	92.0	93.5
Lead Ore	77.8	75.6	75.8	77.9	81.4
Aluminium Ore	43.6	56.3	78.8	84.7	81.2
Timber	68.3	65.6	64.3	64.6	65.3
Zinc Ore	63.3	63.9	61.7	59.0	64.9
Liquefied Petroleum Gas	54.7	57.1	60.5	57.8	58.2
Pulp	18.5	16.5	17.0	19.7	19.2
Iron and Steel Scrap	6.7	4.8	5.2	8.4	7.8

Remarks: Each rate of import was calculated, in principle, from Import/ (Import + Production). The rate for iron and steel scrap was calculated from Import/Consumption. The import of copper ores includes the crude copper produced abroad.
Source: Clean Japan Center

2. Saving landfill space and reducing pollution from landfills

Not only does the landfilling of unprocessed garbage require the greater use of scarce land, but it is also viewed as a significant pollutant. Because of the toxic leachate that often results from the practice, putting raw garbage into landfills is considered environmentally unacceptable and accounts for only 10%-20% of

Japan's total municipal waste stream. Usually, only landfills in rural areas accept unprocessed garbage. According to JICA, reserving landfills for the disposal of inert debris and cemented (or asphalted) incinerator ash facilitates the ultimate reclamation of the landfill for other productive uses. And ash is not allowed to be mixed with raw garbage in landfills. Also, use of ash for cover materials instead of soil is not allowed.

3. Reducing incineration costs and pollution

The Japanese separate recyclables, hazardous materials, and non-combustibles from combustible materials because of substantial economic and environmental benefits to waste incineration. According to the Plastic Waste Management Institute (PWMI):

> Waste separation characterizes collection forms in Japan... [and the] major purpose of waste separation is to prevent the incombustibles from being fed into incinerators, thus promoting volume reduction of the waste to be incinerated ... Moreover, waste separation is expected to encourage citizens to be more conscious of their wastes.

JICA views materials separation as a prerequisite to sizing any community's incinerator. An incinerator's size is established after identifying what should not be burned and what cannot be recycled or otherwise separated. JICA says:

> An important point to be considered in both the treatment and disposal [of waste] is that a system for recovery of resources such as paper, glass, metal, plastics, etc., must be provided in the early stage [of planning an incinerator.]

Thus materials separation helps municipalities avoid the cost of building and operating too large an incinerator, especially now that pollution control is adding to incineration's already high costs. And there's a bonus because large plants are even more difficult to site than small plants, so reducing the size of an incineration plant helps to reduce the social and economic costs. JICA explains:

> Other disposal alternatives must [first] be examined to limit the amount of solid waste [going to an incinerator]...When [alternative] plans are not evaluated in total, the construction of a facility may not provide favorable solutions to solid waste disposal...

In general, as the scale of the project increases, the responsibility for environmental assessment becomes more diffuse and the data available for assessment becomes less specific...[As incineration projects take on a larger] regional scale, the number of groups involved in the decision-making is much increased. This produces problems for those responsible for environmental impact assessments. Unfortunately, these [larger] projects also produce greater quantities of wastes [pollutants] over larger geographical areas.

The environmental benefits of materials separation also include reducing substandard combustion conditions in an incinerator's furnace and keeping out materials which, when burned, could contribute to emissions of heavy metals and other pollutants. By removing non-combustible material from the waste stream, the Japanese reduce the volume of ash residue produced by incineration to about 5% of the material entering the furnace. This saves disposal costs and lengthens the life of scarce landfills.

SEPARATION PROGRAMS

Municipal wastes are separated at the source or at centralized separation centers into metals and metal cans, various grades of paper (corrugated cardboard, newspapers, magazines, office paper, etc.), bottles, glass, textiles, plastics and "wastes difficult to dispose of" (flammable or explosive liquids, toxic chemicals, batteries and, in some cases, plastics.) "Wastes difficult to dispose of" are items more difficult to recycle and those that pose greater pollution problems if they are incinerated or landfilled.

Separation at Source

Households are often required to identify the separated items to facilitate collection. In Hiroshima City, for example, paper and textiles must be bundled with strings "tied crosswise." Metals and bottles must be put out in special, "stout" bags. In agricultural areas, compostable material must be put into "paper bags designated by the town." And according to the PWMI:

At the time of collection, the bags are opened to check the contents. Bags of garbage containing too great an amount of ... [noncompostables] are left without being collected.

Although most Japanese residents self-police their waste sorting, in some cities the name of the disposer must be on the sorted garbage bag so that if a bag is out of compliance with the local sorting regulations, it can be traced. Being identified as a family breaking the rules is a cause for great embarrassment.

Centralized Site Separation

The PWMI states that about 10% of all Japanese cities do not require source separation. Sometimes, "insufficient cooperation ... results ... among citizens." In these cases, a centralized, "processing level" separation of wastes with several techniques, both manual and mechanical, is used.

One method simply involves workers at a processing plant tearing open the polyethylene bags of waste and recovering usable materials. The second, and most popular, puts the bags on a conveyor where one worker opens the bags, a second gathers bottles, a third gathers metals, and so on.

An increasing number of cities, including Matsuyama City (Ehime prefecture), Ueda City (Nagano prefecture), and Matsuzaka City (Mie prefecture), combine manual and mechanical separation. In a few cities laborers at a conveyor sort corrugated cardboard, three color groups of glass (colorless, brown and green) and aluminum. Ferrous scraps are separated both manually and with a magnetic separator. And in a few cities, including Matsuzaka City (Mie prefecture), plastics are separated by an air classifier.

Underscoring the strong Japanese commitment to materials recovery, many cities recover materials at the household level, combined with centralized separation. Sometimes, as in Hiroshima City, processing plants refine the more general separation of items that takes place at the source level. For example, bottles separated from waste by the household are further separated by color at the processing level.

COLLECTION PROGRAMS

Discarded wastes and recoverable materials are collected through house-to-house collection, station collection or fixed-container-based collection. According to JICA, station collection, where waste or recoverable material is deposited at designated stations by individual households, is "most suitable ... in urbanized areas." Station collection is the most common and

Photo: Takuma Co.

Truck washing is practiced at virtually all waste-to-energy plants because Japanese streets are narrow and an idling, filthy truck during collections would upset the local residents.

A trio of buckets for sorting garbage is typically found in Japanese houses and offices.

Public collecting bins, such as this one in Takeyama, are also used to sort waste.

A weighing station collects data on wastes from the municipalities within each region served by a garbage-burning plant.

least costly procedure and usually from 15 to 40 households on a block are grouped as a "station collection" unit.

But station collection is not without its problems. JICA says, "it needs very active public cooperation," is visually "bad" because debris sometimes scatters and "very frequently ... [results in] garbage scattering in the streets." The fixed-container-based collection is a form of station collection but is used almost exclusively at apartment houses.

Most of Japan's reusable materials are recovered from municipal solid wastes in three ways:

1. Voluntary groups collect recyclables either from individual homes or from stations and sell them to a municipal government or to "resource recovery dealers," private enterprises which in turn sell the goods to industry;

2. A municipal government deals only with resource recovery dealers who collect the recyclables themselves;

3. A municipal government carries out all the collection and sales operations by itself.

Collection of wastes is carried out by voluntary groups or private companies in 83% of all municipalities. At the same time, 60% of all municipal governments also operate recycling programs.

Collection by Voluntary Groups

Mr. S. Kato, Director of Solid Waste Management at Japan's Ministry of Health and Welfare (until June, 1987, when he moved to the Environment Agency), says, "we have many, many volunteer groups that collect waste papers, bottles, cans, etc. and sell them for profit, often for children's activities."

Voluntary organizations that collect separate items for re-sale to private industry or intermediate resource recovery dealers are usually registered with the municipal government. This registration enables the local government to assess its own waste collection costs and transport responsibilities. Because it is more costly for the municipal government to collect wastes itself — collection and transport being among the most costly aspects of solid waste management — municipal governments often encourage these voluntary groups to participate in materials recovery by subsidizing them or agreeing to subsidize if the prices of specific commodities drop below an established floor price. To be eligible for a subsidy, the voluntary group is often required to register with the municipality. The types of voluntary groups involved in waste collection in Japan include handicapped groups, PTAs, children's associations, block associations and women's groups.

Examples from three cities illustrate the nature of cooperation among voluntary groups, private resource recovery dealers and municipal governments.

1. In Hiratsuka City (Kanagawa prefecture).

Five categories of recyclables are collected: newspapers, textiles (including leather and rubber), cans, glass bottles and metal scraps. These are collected primarily under the voluntary initiative taken by the town association, local children's groups and PTAs. These groups incorporate public education in their activities to promote cooperation by the local citizenry. Combustible wastes are collected by the municipality according to a regular schedule, and hazardous household materials are collected at a citizen's request, which is the conventional practice in Japan. After collecting the separated items from households or collection bins, the groups deposit the reusable commodities at designated sites, which they are responsible for maintaining. The groups inform the city that the recovered items are available, and the city notifies the local cooperative association of resource recovery dealers which tells one of its members to collect the material. The recovered commodities are weighed, usually in the yard of the resource recovery dealer. Then the city buys the recovered material from the voluntary group that collects it. The purchase of each recovered commodity is guaranteed an 'official' price, which serves as a floor price to encourage the recovery program. The city then sells the material to the dealer at market rates. If the market price is lower than the price paid by the city to the voluntary group, the city, in effect, subsidizes the citizen group.

2. Yokosuka City (also in Kanagawa prefecture).

Uses a "premium payment" program which is a variation of the Hiratsuka municipal subsidy program. Voluntary citizen groups collect waste that is classified into four categories: 1) waste paper, including newspapers, corrugated cardboard and magazines, 2) metals, including ferrous material, tinplates, stoves, fans, water heaters, bicycles, motorcycles, aluminum and copper, 3) glass bottles, and 4) textiles. Other combustible and non-combustible items are collected by the city on a regular schedule, and hazardous household material (including batteries) are collected by the city at the citizen's request.

The group sells the recovered material directly to the resource recovery dealer, who also issues a receipt certifying the amount and type of materials, and the price paid to the group. The citizen group mails this receipt to the city in exchange for a premium equal to one-third the price originally received from the dealer. This 33% premium is designed to increase the cost-effectiveness of the citizen group's materials recovery program.

3. Machida City (Tokyo Prefecture).

Has one of Japan's better recycling programs, which even the communist party, an opposition group, seeks to take credit for. Machida has a population of 320,000, making it somewhat larger than New Haven and somewhat smaller than Boston. To promote recycling, public officials and solid waste management workers go door-to-door at least once a year explaining the purpose of waste separation to residents. The municipality distributes brochures about the benefits of recycling to children in the 3rd and 4th grades. As a result, residents in Machida (with few exceptions) separate their waste into seven classifications: 1) newspapers and magazines; 2) glass bottles; 3) aluminum and steel cans; 4) combustibles, including organic kitchen wastes, light plastics and soiled paper; 5) non-combustibles such as hard plastics, broken glass, scrap metal; 6) poisonous and hazardous materials such as batteries or other items containing mercury or cadmium and 7) bulky wastes such as furniture, discarded bicycles and mattresses.

In Machida, 103 officially registered civic groups, including organizations run by handicapped citizens, carry out the collection of newspapers, glass bottles and metal cans. Resource recovery dealers also collect recyclables. Using a widespread program in Japan, known as "chirigami kokan" (tissue paper exchange), they give residents weekly allotments of tissue paper, napkins and toilet paper in exchange for a week's worth of newspapers. About 50% of the city's glass bottles and steel cans and over 70% of its aluminum cans are recovered.

The municipality collects combustibles, non-combustibles, bulky items and hazardous materials. As in other cities, hazardous materials are collected when requested and stored until safe disposal or recycling methods are found. But they are not incinerated. Bulky items are collected by the municipality once a month and are sent to a recycling cultural center where handicapped citizens repair the items, if possible, for re-sale. Occasionally, 20-tonne batches of organics are composted and used in a greenhouse that is also run by the handicapped group. Repaired bulky items and flowers from the greenhouse are sold to help support rehabilitation programs. In 1985, more than 1,500 items such as bicycles, furniture and books were sold.

Collection by Municipalities and Resource Recovery Dealers

In cities where voluntary groups play a small role in collecting separated material, the municipality licenses resource recovery dealers to collect it. Local governments continue their role in educating the public about waste separation. However in this more simple recovery program wastes are collected less frequently, sometimes once a month. This is a drawback, since one of the benefits of source separation to families in Japan is the frequent collection schedules.

For example, in Kochi City (Kochi prefecture), the government contracts for materials recovery with a co-op of resource recovery dealers. To facilitate the work of the co-op, the city mandates that wastes be divided for purposes of collection into six categories: 1) corrogated cardboard and paper; 2) textiles; 3) metals and metal cans; 4) glass bottles; 5) miscellaneous non-combustibles, including batteries and other hazardous household material; and 6) combustibles, including putrescibles. The municipality collects combustible and non-combustible items and the resource recovery co-op sets a schedule for all the other separated materials. To facilitate the work of the co-op, the city subsidizes the revenue received for the sale of each recovered commodity.

Similar examples are provided by the materials recovery program in Hitachi City (Ibaragi prefecture), and Kashiwa City (Chiba prefecture). In Hitachi City, 25 resource recovery dealers form the Hitachi Resource Recovery Development Cooperative Association. The co-op is divided into three groups, each responsible for one of the three collection districts established in cooperation with the city. Within the three districts are 1,120 collection stations that are each emptied by the co-op only once a month.

Four categories of materials are recovered in this way: 1) paper; 2) textiles; 3) metals and metal cans; 4) glass bottles and other glass items. The city collects combustibles and non-combustibles on a regular schedule and hazardous household materials at the citizen's request.

In Kashiwa City the Kashiwa Resource Recovery Business Cooperative Association is responsible for collecting five categories of recoverable material: 1) paper; 2) textiles; 3) large metal items; 4) glass bottles; and 5) metal cans. Like the Hitachi co-op, Kashiwa collects once a month. The municipality regularly collects combustible and non-combustible items and hazardous household materials upon request. Both the Hitachi and the Kashiwa co-ops receive subsidies from the city twice a year.

Municipal Government as Collector and Dealer

In 17% of Japan's municipalities, no voluntary group or resource recovery dealer exists to collect recoverable material from the waste stream. Although sometimes the municipalities will simply burn or landfill the material, they usually operate a program that separates wastes into six collection groups.

To reduce collection costs the municipality usually establishes collection stations with baskets for each commodity. Sometimes bottles and cans are collected on Bottles Day or Cans/Metals Day. Cities that practice either of these forms of collection include Numazu City (Shizuoka prefecture), Kusatsu City (Shiga prefecture), Fuji City (Shizuoka prefecture), Himeji City (Hyogo prefecture), and Kawanishi City (Hyogo prefecture). Nearly all the collected items are sold. Broken ceramics are landfilled; rubber tires are landfilled or given to any business willing to use them; and plastics are landfilled, incinerated or recycled.

◊

RECYCLING

More than 40 Japanese officials were interviewed by INFORM and all agree that recycling is "the most desirable [waste management] method." They also say that public participation can be easily generated. According to Mr. H. Ogasawara, Foreign Affairs Officer at the Clean Japan Center, "In Japan it is relatively easy to organize the communities to contribute to recycling. For historical reasons we tend to get together to achieve what we need to do for common aims. We easily cooperate."

OVERALL SCALE OF RECYCLING IN JAPAN

Over 3,000 of Japan's 3,255 municipalities have substantial recycling programs going on. JICA found that "As pressures to bring about environmentally sound waste disposal [practices continue to] increase, disposal costs will rise, and [recycling] recovery will become [an even] more attractive alternative." The PWMI disagrees slightly with this because of the costliness and difficulties involved in separating or recycling heterogeneous plastics from the municipal waste stream. The majority of collection for recycling is carried out by civic groups and private industry and few wastes are actually left to be recycled by municipal governments.

Recycling data in Japan indicate how much of a commodity or primary material is produced, how much of it is returned and how many times it is reused (in the case of bottles), or how much of it is used in new production. However, government recycling data cover only about 60% of Japan's 3,255 municipalities. When government documents state that "municipal separation programs" exist in 60% of Japan's municipalities, this percentage refers only to recycling programs operated by municipal governments. The documents do not take into account the privately or voluntarily operated recycling programs that exist in 83% of the municipalities.

The Japanese do not maintain aggregate recycling data because they consider reusable materials to be resources, not wastes. They do not make their way into Japan's waste stream and, therefore, are not counted as wastes by the municipalities. As indicated later in this chapter, the data that do exist are commodity-specific and the rates are high:

> 66% of bottles are used an average of 3 times

> 42% of glass bottles are made from cullet

> 51% of paper is recovered

> 55% of steel is recovered

In addition to the usual newspaper, magazine, glass and can recycling programs, other materials are sometimes recycled: Some municipalities sort plastic bottles for recycling; 27% of the nation's button-shaped mercury batteries are recycled; and 20% of Japan's fabrics are recycled. Bulky wastes, such as furniture, bicycles, lumber and appliances, even refrigerators and washing machines are often collected separately and sent to shredding facilities adjacent to incinerators. After shredding, ferrous metal is recovered and sometimes aluminum and glass as well.

The overall recycling rate in Japan is simply not known. However, based on estimates that in the U.S., paper, metal and glass alone can easily account for 45-60% of the waste stream, it is likely that in Japan, with all categories of waste included, the overall national figure for recycling is about 50%.

This was confirmed by Dr. M. Tanaka of the Institute of Public Health. After cautioning that no classification of this sort exists, Dr. Tanaka estimated that "perhaps" 50% of Japan's "waste" is recycled if Western terms are applied. Since the rate of recycling does vary by municipality in Japan, there are instances where local recycling exceeds or falls short of the national average. The City of Machida, for example, may be recycling as much as two-thirds of its waste, while Osaka may be recycling less than 50% overall.

MATERIAL RECYCLED AND RECOVERED

Waste Paper

Waste paper recycling has been going on in Japan since at least the Edo period (1600-1868) and the Japanese even have two words for paper, one referring to "new paper making" and one referring to "re-paper making," using waste paper. Because Japan has recycled waste paper since before waste incinerators were developed, it is safe to assume that recycling is motivated by the need to conserve resources rather than to reduce loads on incinerators. As indicated in Table 6 in Chapter IV, Japan's timber imports averaged about 65% between 1980 and 1984, while pulp imports averaged 18% for those same years. Saving energy is another reason waste paper is actively collected. According to the Clean Japan Center, collecting used paper plays

an important role in saving energy and reducing waste. As one tonne of waste paper is said to be equivalent to 20 green trees (about 14cm in diameter and 8m in height), such activities clearly play an important role in the conservation of resources. The consumption of paper and cardboard in 1981 was about 17 million tonnes. With the recovery of 8 million tonnes of waste paper, the recovery ratio was 47 percent. The energy consumption when using waste paper is estimated to be about 2/3 of that when using pulp.

The recovery rate of waste paper more than doubled between 1955 and 1984, from just over 20% percent to 50.7% (see Chart 3). Figures for 1983 indicate that 18.6 million tonnes of paper were consumed, while about 9.1 million tonnes were collected for reuse. Data for paper consumption and collection for 1980-1983 are noted in Table 7.

TABLE 7 Recovery Rates for Wastepaper

	1980	1981	1982	1983
Amount of paper consumed (1,000t)	16,816	17,089	17,434	18,561
Amount of waste paper collected (1,000t)	8,017	8,125	8,460	9,148
Recovery rate (%)	47.7	47.5	48.5	49.3

Source: Ministry of Health and Welfare

CHART 3   Changes in Recovery Rate and Use of Waste Paper

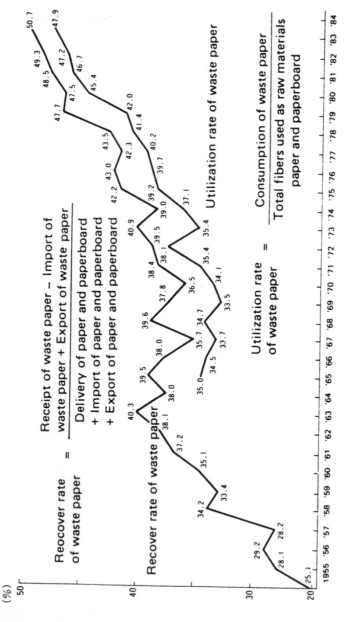

Source: Clean Japan Center

Because of Japan's high dependence on imports, the Clean Japan Center argues that "further promotion of waste paper recovery will continue to be an important task in the future." It estimates that additional waste paper can be recovered and, consequently, "in theory...the recovery ratio can be raised...up to 60%. The remaining 40% is [deemed] unrecoverable, including tissues and toilet paper." As recently as 1983, Japan's waste paper recovery rate was more than twice that of the U.S.

Glass Cullet

Cullet is crushed glass used as the raw material for producing new glass. In 1984, 42% of Japanese glass bottles were made from cullet. Although temperatures as high as 1,600°C (2,900°F) are needed to make new glass bottles from cullet, the Japanese estimate the following savings in fuel and the three main raw materials for making glass when cullet is used (see Table 8).

TABLE 8 Savings in Making Glass per Tonne of Cullet Used

	(at 1981 costs and yen values)
Soda	217 kg (¥11,800)
Silica sand	734 kg (¥4,800)
Lime	166 kg (¥800)
Heavy oil	37.1 liters (¥2,400)

Source: Clean Japan Center

TABLE 9 Output of Bottles and Quantity of Cullet Used

	1980	1981	1982	1983	1984
Output of bottles(1000t)	1,957	1,740	1,907	2,147	2,240
Quantity of cullet used (1000t)	695	678	801	888	949
Percentage of cullet used(%)	35.5	39.0	42.0	41.3	42.4

Source: Japan Bottlers Association

These savings in raw materials are produced by the present 42% ratio of cullet to raw materials (see Table 9). However, the Japanese hope to increase the ratio of cullet in glass. According to the Clean Japan Center:

Though there are problem areas such as the difficulty of extracting air bubbles from [glass with a] high ratio of cullet, it is said to be possible to mix in cullet up to 60 percent. Special bottles of 100 percent are also being developed...It is thus important to place not only returnable·bottles but as many one-way bottles [cosmetics, medicines, etc.] as possible on a recycling route.

Although an increase in the rate of cullet used in glass production can yield only marginal energy savings, the Japanese seek to achieve them: According to the Clean Japan Center, "If the cullet-use rate is increased by 1%, the energy required for melting glass can be reduced by about 0.15%."

TABLE 10 Proportion of Complete Bottles Reused (FY 1983)

Items	Quantity used (1,000s)	Quantity collected (1,000s)	Percentage of bottles reused	Average number of times used
2-liter sake bottle	7,830	7,440	95	20.0
Beer bottle	608,970	578,520	95	20.0
Juice bottle	274,000	250,890	91	11.1
1.8-liter sake bottle	122,250	102,690	84	6.3
Soda bottle	332,000	258,400	78	4.5
Yogurt	11,300	2,030	18	1.2
Soy sauce	32,500	5,530	17	1.2
Domestic whiskey bottle	61,810	8,040	13	1.1
Middle & small size sake bottles	76,400	8,300	11	1.1
Medicinal drink bottle	281,500	0	0	1.0
Instant coffee bottle	31,980	0	0	1.0
Imported whiskey bottle	7,800	0	0	1.0
Total	1,848,440	1,221,840	66	2.9

Source: Japan Bottlers Association

Refillable Bottles

The Japan Bottlers Association in cooperation with the Japanese government maintains precise data on glass collection, bottle-by-bottle, which indicate great variations in bottle collections and reuse. For example, while 95% of all new beer and two liter sake bottles were collected and reused an average of 20 times, only 17% of soy sauce bottles were collected and reused only 1.2 times. No instant coffee bottles are reused, although they may be made with recycled cullet (See Table 10). About 66% of all bottles in Japan are collected and reused an average of three times. One-way bottles, such as those containing medicine or cosmetics, are landfilled, not reused, and kept out of incinerators because they are not combustible. Glass recovery varies by region and locality (See Table 11).

The routes travelled by bottles from consumer to bottling plant or bottle manufacturer (or, in some cases, a landfill) are illustrated in Chart 4.

TABLE 11 Glass Recovery by Region (1983)

Regions	Number of Cities Surveyed	Cities Responding Number	Percent	Percent Recovered Empty Bottles	Cullet
Hokkaido	32	23	(71.9)	33.3	0.0
Tohoku	64	49	(76.6)	58.8	29.4
Kanto	152	97	(63.8)	68.7	62.7
Hokuriku	44	34	(77.3)	11.1	0.0
Chubu/ Tokai	101	70	(69.3)	53.7	65.9
Kinki	86	57	(66.3)	37.9	65.5
Chugoku	48	34	(70.8)	55.6	33.3
Shikoku	30	20	(66.7)	55.6	44.4
Kyushu/ Okinawa	90	60	(66.7)	44.4	22.2
Total	647	444	(68.6)	54.0	51.5

Source: Japan Bottlers Association

CHART 4 Basic Recovery Routes for Bottles and Cullet

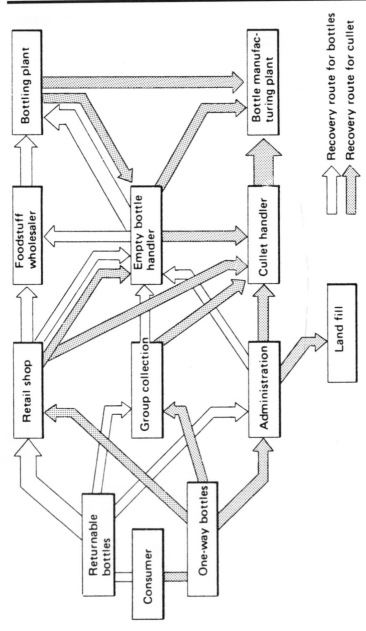

Source: Clean Japan Center

A street in Kyoto where cardboard is collected for recycling.

Bulky wastes, such as bicycles and furniture, are sometimes repaired and resold for profit.

Battery displays at recycling cultural centers teach children and other citizens about the pollution that improper disposal can cause.

Metals: Aluminum and Steel

Because aluminum is lightweight and more corrosion resistant than steel, its use in beer and soda cans has been increasing in Japan from 33 million cans in 1971 to 1.74 billion cans in 1981 (See Chart 5). The Clean Japan Center estimates that using recycled aluminum cans to produce a new aluminum product requires only 1/26 (96% less) of the energy needed than when bauxite, the raw material for aluminum, is used. For this reason aluminum is described by Japanese recyclers as the "leading runner for energy savings."

Although percentages for the recovery of aluminum beverage cans vary by region, from a low of 33.3% in Hokkaido to 88.9% in Shikoku, currently 40.6% of aluminum cans are recovered nationwide. The recovery in 1981 of 11,000 tonnes is estimated to have saved about 190 million kwh, equivalent to the annual energy consumption of over 80,000 Japanese households.

Significant energy savings are also achieved when steel cans are recovered and used in new production, although the 65% reduction in needed energy is not as dramatic as the 96% reduction from reusing aluminum. Despite this, more steel cans are recovered than aluminum cans, mainly because magnets can help recover steel but cannot be used to recover aluminum.

CHART 5 Changes in Can Production (Beer and Carbonated Drinks)

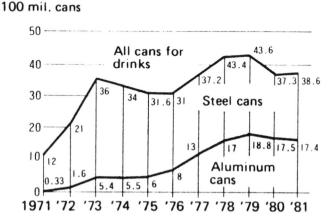

Source: Clean Japan Center

TABLE 12 Production and Collection of Steel Products

	1980	1981	1982	1983
Amount produced (1,000t)	65,568	56,592	53,903	50,816
Amount collected (1,000t)	26,596	27,084	26,499	27,695
Recovery rate	40.6%	47.9%	49.2%	54.5%

Sources: International Trade and Industry Statistics Survey and Japan Association of Iron Scrap Industry

In 1981 over 40% of all steel cans were recovered (394,000 tonnes), for an estimated saving of about 470 million kwh, which is equal to the annual electrical consumption of over 200,000 households. More than half of all types and forms of steel is now recovered in Japan (See Table 12).

Plastics

Japan ranks ninth in per capita plastics consumption (68.3 kg/y) among the non-Communist, industrialized states, while the U.S. ranks sixth (86.4 kg/y) (See Chart 6). About 24% of Japan's plastic consumption is associated with packaging, which is less than the amount of plastics used in packaging by any other non-communist nation with the exception of West Germany and Switzerland (See Table 13).

Plastics now account for about 5 to 8% of Japan's municipal waste, and the percentage is increasing. During the 1970s the increasing amount of plastics in the municipal waste stream raised two particular concerns among the Japanese public.

First, older waste incinerators could not handle the substantially higher caloric value of wastes having high percentages of plastics. Plastics have a heat value ranging from 4,500 to 11,000 Kcal/kg, which raises the average heat value of municipal waste to about 1,500-1,700 Kcal/kg. The designed maximum caloric values for the hundreds of municipal incinerators built before the 1970s in Japan was below 1,500 Kcal/kg (See Chart 7).

Second, plastics, especially those with a high percentage of PVC resin, have been identified as a major source of incinerators' HCl emissions, an acid gas of great public concern. Although Japan's Plastic Waste Management Institute disagrees, JICA and hundreds of Japanese cities have labeled plastics as "unsuitable matter for combustion," or a "waste difficult to be disposed of." Because of these reasons, various actions have been taken throughout Japan to reduce the amount of plastics going to incinerators. For example, a 1982 survey covering Japan's 3,255 mu-

CHART 6 Per Capita Plastic Consumption in Major Countries

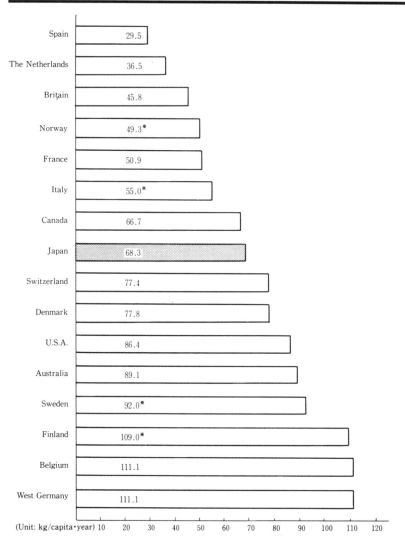

Country	kg/capita·year
Spain	29.5
The Netherlands	36.5
Britain	45.8
Norway	49.3*
France	50.9
Italy	55.0*
Canada	66.7
Japan	68.3
Switzerland	77.4
Denmark	77.8
U.S.A.	86.4
Australia	89.1
Sweden	92.0*
Finland	109.0*
Belgium	111.1
West Germany	111.1

(Unit: kg/capita·year)

Source: PWMI

Table 13 Plastics Consumption by Consuming Sector in Major Countries (1981) Unit: %

Consuming Sector	Building	Packaging	Electric/electronics	Transport	Furniture	Agriculture	Play articles/toys	Household articles	Footwear/cloth	Machine parts	Liquid Products (adhesives/paints)	Others
Australia	21	31	10	4	11	4	1	3	2	7	4	2
Belgium	18	30	25	5	5	3	--	4	5	5	--	--
Canada	22	33.5	4.2	9.6	7.4	4.2	4.8	3.6	1.3	4.6	4.8	4.8
Chili	18	30	5	4	6	4	4	8	2	--	--	19
Denmark	15-20	26-30				50~60						
West Germany	25	21	15	7	5	4	--	2.5	--	--	10	10.5
France	18.4	35.1	7.6	7.5	3.7	4.8	--	5	--	2.6	--	15.3
Italy	10.5	31	9.5	5.7	5	4	7.5	5.5	0.7	1	14.6	5
Japan	13.1	23.7	13.1	7.5	1.1	3	1.3	8.4	0.3	1.6	14.4	12.5
The Netherlands	11	47	4	1	--	--	1	12	1	2	--	21
South Africa	12	34	10	7	4	3	3	2	7	4	5	9
Spain	12	30	10	5	6	5	2	5	1	6	6	12
Sweden	18	26	13	5	4	1	--	3	--	16	--	14.0
Switzerland	26	21	12	4.5	4	5.5	--	5	--	7.5	--	14.5
Britain	20	34	10	5	7	2	4	2.5	1	2	--	12.5
U.S.A.	19.6	28.3	7.2	4.3	4.5	--	--	9.9	--	1.1	7	18.1

Source: PWMI

CHART 7 Changes in Caloric Values of Municipal Wastes

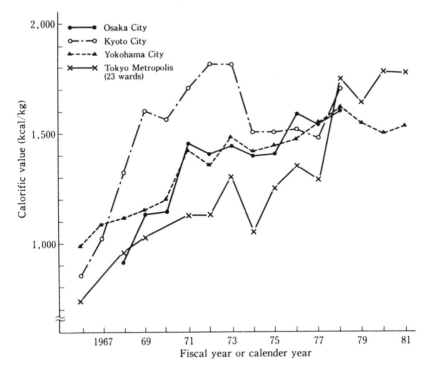

Source: PWMI

nicipalities, indicated that about half separated plastics from combustible waste for direct landfilling, while 12 cities separately collected plastics for single fuel incineration.

The concern about the increasing use of plastics has given rise to actions to reduce their use. For example, a 1970s National Municipal Sanitation Conference adopted a resolution opposing the introduction of plastic milk containers. And at the 1985 International Exposition of Science and Technology (EXPO) in Tsukuba (Ibaraki prefecture), plastic throw-away containers were banned, a symbolic action previously undertaken at Osaka's EXPO '70.

As with all municipal waste, municipalities decide how to handle plastics (See Table 14). Although about half the municipalities remove plastics from combustible waste, the amount

TABLE14 Ten Examples of Plastic Recycling Activities by Municipalities

Municipality (Prefecture)	Plastics subject to resource recovery	Roles of municipal office	Amount of recovered resources/period	Remarks
Hitachi (Ibaragi)	Large products of expanded styrene and rigid plastics	To separate plastics from the other combustibles and bulky wastes at the public cleansing center. Expanded styrene is melted and solidified with a machine, then sold.	36 tons (expanded styrene), 5 tons (rigid plastics)/ FY1980	Initiated on April 1, 1979
Fuchu (Tokyo)	Large plastic products (barrels, detergent containers, buckets, etc.)	To separate plastics from the incombustibles at an incombustibles disposal facility and sell them.	40.82 tons/ FY1980	Initiated on April 1, 1979
Fussa (Tokyo)	Plastic containers	To separate plastics from the incombustibles at an incombustible disposal facility and hand them over to contractors.	39.84 tons/ FY1980	Initiated on April 1, 1977
Kamakura (Kanagawa)	Mainly polyolefin resins	To separate plastics from the incombustibles and hand mainly polyolefin resins over to plastics reclamation operators free of charge.	17.19 tons July 1981- Jan. 1982	Initiated in July 1981 on an experimental scale
Fuso (Aichi)	Detergent containers vinyl bags, plastic toys, food containers, plastic lids, etc.	To classify the collected plastics into various groups and sell them.	26.6 tons/ April 1981 - Febl 1982	Initiated in Nov. 1980

Kusatu (Shiga)	Mainly light-weight plastics, while all the kinds of plastics are collected.	To crush or pulverize plastics and melt and mold them after the washing/classification/dewatering/drying processes	760.07 tons/ FY 1980	Molded products include flower pots and stakes.
Himeji (Hyogo)	Food trays, packaging materials, detergent containers, plastic toys, bags, shoes, expanded styrene, etc.	To classify plastics into polymer plastics and expanded styrene with a special plastics separator. Both groups are sold, the former after being crushed and the latter after being melted (crushing/melting operations by contractors).		Used as raw materials to produce stakes, plates, etc.
Kawanishi (Hyogo)	Overall plastics.	To dispose of the collected plastic wastes.		
Nishiwaki Takino Kurodasha (Hyogo)	Plastics other than expanded styrene.	To separate plastics from the incombustibles manually, crush them, and sell them.	28 tons/ FY1980	Intiated in fiscal 1979
Ohno Asaike Ogata Kiyokawa (Ohita)	Food containers, detergent containers, buckets, plastic parts of TVs, washing machines, etc.	To manually separate plastics from the resource-containing wastes, which are washed and dried by the sun. To classify the treated plastics by resin and color. To produce pellets or molded products, which are sold.		Initiated in Dec. 1981

Source: PWMI

TABLE 15   A Survey of Plastics Recycling by Region (1983)

Regions	Number of Cities Surveyed	Cities Responding		Percent of Plastics Recycled
		Number	Percent	
Hokkaido	32	23	(71.9)	0.0
Tohoku	64	49	(76.6)	0.0
Kanto	152	97	(63.8)	6.0
Hokuriku	44	34	(77.3)	0.0
Chubu/Tokai	101	70	(69.3)	0.0
Kinki	86	57	(66.3)	10.3
Chugoku	48	34	(70.8)	11.1
Shikoku	30	20	(66.7)	11.1
Kyushu/Okinawa	90	60	(66.7)	5.6
Total	647	444	(68.6)	5.0

Source: Ministry of Health and Welfare

(volume or weight) of material being recovered is uncertain. Under 5% of the plastics in the municipal waste streams are actually recycled, although in the Kinki, Chugoku and Shikoku regions as much as 11% is recycled (See Table 15). Plastic stakes for fences, tents and playground construction are Japan's leading reclaimed plastics product.

Because industrial processing of plastics tends to generate more homogeneous plastic resin than municipal solid waste, 31% is recycled from Japan's industrial waste. According to the PWMI, reclamation of plastics for use as raw materials,

is applicable only when plastic wastes meet some preconditions...(1) they [must] consist of a homogeneous resin, (2) they [must be] available in large quantities, and (3) they [must be] free from inclusion of foreign matter and contamination.

This is a formidable task, especially given the variety of plastic resins widely used in Japan (See Chart 8). Nevertheless, government and industry funds have been allocated to explore this recycling option.

Batteries

Two collection programs for recovering batteries exist throughout Japan. Cylinder-shaped (manganese and alkaline-manganese) batteries are collected by municipalities. About 73% (2,374) of Japan's 3,255 municipalities are now involved, incor-

CHART 8 Output of Plastics by Resin

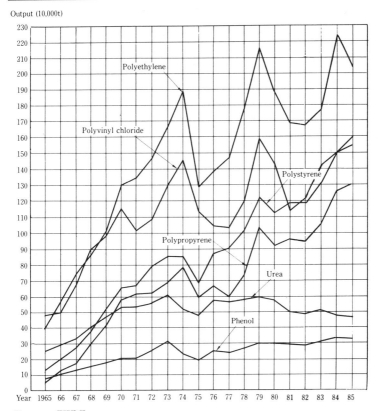

Output (10,000t)

Year 1965 66 67 68 69 70 71 72 73 74 75 76 77 78 79 80 81 82 83 84 85

Source: PWMI

porating more than 86 million people or about 72% of the total population. Button-shaped mercury batteries are collected by battery manufacturers on a non-municipality basis through retail stores, involving 100% of the population. According to Mr. Kato, of the Ministry of Health and Welfare, "although only mercury batteries are targeted [by manufacturers] to be separately collected, all kinds of button-shaped batteries (alkaline-manganese, silver oxide and lithium batteries) are separately collected because they are not distinguished easily when discharged."

Interestingly, despite the widespread participation of the Japanese population in the collection of batteries, only 9% by weight of the cylinder-shaped batteries are recovered. Similarly, despite a program involving all of Japan's population, only 27%

TABLE 16 Battery Separation & Recycling in Japan

	Cylinder-shaped[1] 6/85 - 5/86	Button-shaped[2] 1/85 - 12/85
Batteries consumed		
Number	1,585,000,000	12,472,000
Weight (tonnes)	68,155	NA
Batteries separated		
Number	NA	3,364,000
Weight (tonnes)	6,300	NA
Percent	9	27
Municipalities involved		
Number	2,374	Collected by manufacturers
Percent	73	in all municipalities
Population involved		
Number	86,000,000	119,000,000
Percent	72	100
Disposition of separated batteries	By municipality: −548 send batteries to a mercury-refining factory −1,762 send to temporary storage −47 use concrete solidification disposal	All the manufacturers send to a mercury-refining factory in Rubeshibe, Hokkaido

1. Includes both manganese and alkaline-manganese batteries
2. Although only mercury batteries are targeted, all button-shaped batteries (including alkaline-manganese, silver oxide, and lithium) are collected because they are not easily distinguished.
Source: Ministry of Health and Welfare.

by number of the button shaped batteries are removed (See Table 16).

The reason for this low recovery rate, despite the high participation rate, is that this program is newly instituted. According to Mr. H. Ogasawara of the Clean Japan Center, "it requires more years for this system to be accepted by the population."

All of the collected button-shaped batteries go to the Itomuka Mercury Refinery in Rubeshibe City, Hokkaido. This refinery, which is a demonstration project first set up in 1984 and subsidized by the Ministry of International Trade and Industry, also accepts dry cells, fluorescent lamps, thermometers and mirrors.

Of the municipalities that collect cylinder-shaped batteries, 548 send the batteries to the Itomuka Refinery, 47 mix the batteries into concrete and then landfill the material and 1,762 store the batteries until safe and economical reprocessing or disposal options become available.

It costs ¥75,000 to treat and dispose of 1 tonne of the collected cylinder-shaped batteries at the factory and costs ¥6,000-¥33,000 to send them there, depending on the distances involved. These costs are all borne by the municipalities. According to Mr. Kato, the "separate collection, treatment and disposal of the button-shaped batteries are economically self sustaining because they contain silver oxide," which is valuable.

The removal of batteries from the waste stream going to incinerators or landfills is not based on health-effect data. Mercury emissions from incinerators were tested in 1984 and the measured levels were determined not to be a health threat. Japan collects batteries for two reasons. First, as with most other recovered materials, to conserve resources. Commenting on MITIs Itomuka Mercury Refinery, Mr. H. Ogasawara of the Clean Japan Center said "it is necessary not only for technical development, but [out of] economic consideration [as well] to promote waste recycling." The second reason is the public's concern about hazardous mercury emissions, despite the absence of substantiating health effect data. Mr. Kato of the MHW explains:

The result of environmental monitoring tells us that treatment [incineration] of waste batteries doesn't cause environmental pollution even when they are treated [incinerated] together with other garbage.

But still, along with the increase of alkaline manganese batteries (which are higher in mercury content than conventional manganese batteries), and because of the development of environmental concerns, people [are]...uneasy about potential environmental pollution by mercury. Because of this...apprehension, many municipalities found it difficult to smoothly carry out waste management practices such as constructing incinerators and landfill sites.

Therefore, many municipalities voluntarily started to separately collect waste (cylinder-shaped) batteries from other garbage in order to secure smooth implementation of waste management by meeting the people's needs for much safer, better and more reliable waste management.

The Ministry of Health and Welfare concluded in June 1985 as follows:

1) There is and will be no environmental pollution by waste battery treatment [incineration].

2) Among the measures to meet people's needs for a better, safer and more comfortable environment, the best is [that] battery manufactures [should] reduce the mercury content of alkaline-manganese batteries.

3) Each municipality should carefully judge whether there is any need to separate and collect the batteries based on the above mentioned counter-measures [taken] by the manufacturers.

4) A new option (i.e. mass treatment at mercury refining factories) should be established in order to rationally treat the collected (cylinder-shaped batteries which ... [have been] already collected and are [now] stored.

In 1983, the MHW negotiated an agreement with battery manufacturers to reduce the mercury content of the alkaline-manganese batteries to one-sixth of 1983's content by the fall of 1987.

Tires

Data on the number of tires recovered in Japan are not readily available, nor is information available on the collection processes involved. But it is apparent from the increasing use of automobiles that the number of discarded tires has been increasing annually. The number of tires discarded in 1978 (the only year for which INFORM has data) was about 53 million, or about 550,000 tonnes in weight.

Out of a concern about the increasing burden that discarded tires place on municipal solid waste management programs, the Clean Japan Center constructed a recycling plant in 1970 at Ako (Hyogo prefecture). The plant is designed to recover fuel gas, oil and high quality carbides (See Table 17). It can handle 7,000 tonnes of discarded tires each year.

TABLE 17 Ako Thermal Decomposition Plant

	Kilograms/hour	Tonnes/year
Tire treatment capacity	1,100	7,000
Resources Recovered		
Gas	66	420
Low grade oil	160	1,015
High grade oil	445	2,835
Iron	55	350
Refined carbon	374	2,380

Source: Clean Japan Center

Composting

Although the number of high-rate composting facilities in Japan has increased from 15 in 1978 to 23 in 1983, their total capacity declined from a peak of 700 t/d in 1978 to 389 t/d in 1983 (See Table 18). Actual amounts composted in 1983 were 0.2% (211 t/d) of all Japanese wastes. This decline has been due in part to the increased use of chemical fertilizers in place of compost. It is also due to the difficulty in making, transporting and using the compost. However, as the dangers inherent in chemical fertilizers become more apparent, it is not unreasonable to expect that composting will gradually increase.

TABLE 18 Composting Plants in Japan

End of FY	Number of plants	Capacity (tonnes/day)
1978	15	700
1979	15	630
1980	12	500
1981	15	412
1982	20	434
1983	23	389

Source: "Disposal of Waste in Japan," Ministry of Health and Welfare (FY 1983 edition)

According to the PWMI:

In Japan, composting has been losing its popularity ... in recent years. This ... diminishing demand for compost [is] a result of [the] growing use of chemical fertilizers which feature quick effects with a limited amount of use. Recently, however, organic fertilizers have been regaining attraction as it is now revealed that continuous use of chemical fertilizers can cause such problems as acidification of soil and deteriorating fertility...

Most of the composting in Japan is in agricultural areas. Many of these areas still rely on landfills and have not yet incorporated an incinerator into their solid waste management program. Consequently, in rural towns such as Usuda (Nagano prefecture), where farming families account for half the total population, waste is separated only into two groups: 1) organic, compostable material and 2) everything else, which is landfilled.

Composting in Japan is a centralized operation and municipalities own and operate the composters. The PWMI describes the process generally used in rural areas as follows:

Townspeople are required to separate their wastes into two groups: [compostable] garbage and wastes to be landfilled. Each group must be packed in paper bags designated by the town. At the time of collection the bags are opened to check the contents. Bags of garbage containing too great an amount of [non-compostables] are left without being collected. Garbage forwarded to a compost production center comes from not only general households but the business sector, including food processing plants and greengrocers' shops. At the [compost production] center wood chips are first added to the garbage to reduce its moisture content. Then the garbage is crushed. Left in a primary fermentation tank for about a week, the garbage is screened and moved to a secondary fermentation tank. With livestock raw sewage sprayed over it, the garbage is stored in the open air for a few months. When it turns to compost, it is returned to townspeople.

Not all farming in Japan is limited to rural areas and many non-rural cities also use high rate composters for local farming, including Nagasaki City (Nagasaki prefecture), Machida (Tokyo prefecture), and Toyohashi City (Aichi prefecture). Within the more urban settings -- and Japan is mostly an urbanized nation – the procedures for composting vary. In Nagasaki, for example, the city operates only the primary fermentation process, in a rotating fermentation tank. The compost is used for trees planted in public spaces or is sold to local farmers who maintain the compost through its secondary fermentation process.   In Toyohashi City and Machida City the composting takes place in a resource recovery cultural center. In Toyohashi the compost is returned to farmers, while in Machida it is used by handicapped groups in their greenhouse to make saleable botanical arrangements.

High rate composting is not without its adverse environmental effects, especially when, as is the case of Toyohashi City, organic wastes are mixed with dewatered sludge to form compost. According to JICA, the adverse impacts include odors and, most seriously, "contamination of soil and underground water by the harmful components of the original refuse." Paper often winds up in compost, and because it can contain phenols, cadmium, lead and mercury, compost can cause heavy metal and other forms of pollution unless it is carefully maintained. This has also contributed to the increasing reluctance by farmers to use municipally-produced compost. Because of this possibility, the Japan Ministry of Agriculture, Forestry and Fisheries has established the following quality standards for compost:

TABLE 19 Compost Standards

Name of substance	Standard value
Arsenic	50 ppm or less for total analysis (dry base) 1.5 ppm or less for leachate test
Cadmium	5 ppm or less for total analysis (dry base) 0.3 ppm or less for leachate test
Mercury	2 ppm or less for total analysis, 0.005 ppm or less for leachate test
Lead	3 ppm or less for leachate test
Organic phosphorus	1 ppm or less for leachate test
Hexavalent chromium	1.5 ppm or less for leachate test
Cyanide	1 ppm or less for leachate test
PCB	0.003 ppm or less for leachate test

Remarks: 1) C/N ratio shall be 20 or less. 2) Total nitrogen shall be 2% or more per dry base. 3) Reducing sugar rate shall be 35% or less.
Source: JICA

Public Education and Cooperation

Of all the elements which contribute to Japan's extraordinary materials separation and recycling success (see Box, next page),

---

Five Reasons for Japan's Recycling Success

1. Over a century of experience

2. Dependence on imported raw materials

3. Dedication to pollution prevention

4. Government support of recycling

5. Public education and cooperation

---

perhaps the most effective tool is the widespread public education programs.

Municipalities and the Clean Japan Center print brochures explaining garbage as an environmental problem and an industrial resource and telling about the benefits of materials separation and recycling [See Appendix 2). These brochures are circulated widely and are included in the curriculum of 3rd and 4th grade students. Students are routinely taken on tours of garbage-burning plants and told about the benefits of recycling. However, according to Mr. Ogasawara of the Clean Japan Center, "Children, most of the time, learn [how to separate garbage] by experience, at home." In some cases -- such as in Machida, which could have as much as a 65% recycling rate, public officials and solid waste management workers go door-to-door at least once a year explaining the waste disposal system and the benefits of waste separation and recycling.

According to JICA, the following 12 "cost effective" items should be included in all public education materials about waste:

1. The economic and health benefits of a cleaner city

2. What is expected of the resident as part of the cooperative effort

3. How a resident can obtain information on the system or make comments or complaints

4. Organization of the system

5. What improvements are planned

6. The annual budget

7. How many workers are involved

8. What services are provided

9. Designation of special pick-ups for recyclables

10. Arrangements for special collection of bulky wastes or garden wastes

11. Requirements for storage container placement and removal for pick-up

12. Schedule of collection pick-up.

Many Japanese municipalities also attach violation tags to the garbage bags so that the resident learns what mistake has been made. Fortunately Japan is an extraordinarily cohesive and responsive society when carrying out government policies, as illustrated by INFORM's conversation with one Japanese citizen (See Box).

The Limits of Recycling

According to JICA, "most of the materials in municipal waste are recoverable in one form or another," and yet, despite the fact that so cohesive a society is being widely informed about the

---

Mari   Hatano

Mari Hatano lives in Shiga Prefecture near the city of Kyoto. She must separate her household waste into six categories, including four put into kitchen bags: paper and kitchen waste, plastics and incombustibles, glass, and cans. These items are collected on different days of the week. Pickup is not at her home but at a centralized collection point approximately one block away. Newspapers and magazines are collected door-to-door by a citizens group and bulky waste is picked up at her home once each month. Mari says that separating waste is extra work and she doesn't particularly enjoy doing it, especially after working all day and coming home late — but she does it nevertheless. There are no fines or penalties for not complying with the rules but people, in general, do. As Mari told INFORM, she "wants to see her prefecture clean." She wouldn't even think of not doing it.

benefits and need to recycle, and although the government often financially supports such efforts, only about half of all wastes in Japan are recycled.

Mr. H. Ogasawara, Foreign Affairs Officer at the Clean Japan Center, does not believe Japan can do much better than recycle half its wastes and takes issue with those who argue for total recycling (85% or better):

> Total recycling? No, that's impossible. People tend to believe in recycling too much. There is only so much you can do. In Japan, we believe we have just about reached the limits of what recycling can do...The social aspects are key, much more important than the technological aspects of recycling.

A similar position is maintained by Mr. K. Onogawa, Deputy Director for Water Quality at Japan's Environment Agency, who oversees Japan's landfills, and by Mr. S. Kato, Director of Solid Waste Management at the Ministry of Health and Welfare, who has stated:

> It is easy to get people to recycle. But total recycling? A few people are saying that, but it is not possible. They are claiming that we should increase recycling and reuse. My response, and the response of the government, is yes, we can still reduce the amount of waste by 3% percent, maybe.

VI

INCINERATION

Incineration is regarded by the Japanese as the most sanitary method to treat wastes. According to JICA:

The basic purposes of interim treatment [incineration] are: 1) to reduce volume, 2) to biologically stabilize [waste] for sanitary handling, and 3) to neutralize hazardous substances...

Japan has constructed 1,915 waste incinerators in sizes up to 1,800 tonnes/day. These facilities process 68% of post-recycling waste. This represents the greatest commitment to waste-incineration of any country and complements Japan's extraordinary recycling efforts, which appear to be the most comprehensive and successful in the world. In 1983, the capacity of all incinerators was 153,303 tonnes/day, with an average throughput of 75,022 tonnes/day. While most of the incinerators are of the small batch or intermittent type, the bulk of incineration capacity is in the larger, newer fully continuous waste-to-energy facilities. There were 361 of these in 1983 (See Table 20).

TABLE 20 Solid Waste Incineration Facilities in Japan

(Number, capacity, and tonnes/day)						
	Fully continuous*		Other types**		Total	
Year	No.	Cap. t/d	No.	Cap. t/d	No.	Cap. t/d
1978	344	94,964	1,681	43,803	2,025	138,767
1979	341	96,396	1,643	48,177	1,984	144,573
1980	357	98,914	1,642	47,980	1,999	146,894
1981	370	101,399	1,613	46,981	1,983	148,380
1982	360	103,479	1,582	47,874	1,942	151,353
1983	361	105,152	1,554	48,151	1,915	153,303

* Fully continuous are all waste-to-energy.
** Other types include batch (manual and mechanical) and intermittent.
Source: *Solid Waste Management and Night Soil Treatment (I)*, JICA, and *Quality of the Environment in Japan* 1985, EA.

Most Japanese plants process garbage as received ("mass burn"), but, as noted previously, waste is widely presorted with much non-combustible material removed. Even the larger Japanese plants are generally smaller than many of those planned (or operating) in the U.S. The eight facilities visited by INFORM were chosen because they were built by four different vendors and because of their advanced technological features. They are in the 195 to 1,200 tonnes/day range (See Table 21). The largest plant in Japan is a 1,800 tonnes/day Tokyo plant which is half the size of the largest U.S. waste incineration complex. It is very common for incinerators to have shredders for processing bulky waste, which accounts for as little as 1 to 5% of the waste going to incinerators. After metals are recovered from the bulky waste, the combustible fraction is separated and incinerated. Co-incineration of dewatered sewage sludge with waste is not common but is gaining in popularity. The Sohka plant visited by INFORM burns 21.5 tonnes/day of dewatered sewage sludge delivered by underground conveyor from an adjacent sewage treatment facility. Incinerator flue gases are used to dry the sludge before combustion.

TABLE 21 Ownership and Scale of Plants Visited

Plant (Manufacturer)	Ownership/ Operation	Scale
Machida City (IHI)	Public/Public	3 furnaces/ 150 t/d each
Kumamoto City/ Western Plant (Takuma)	Public/Private	2 furnaces/ 225 t/d each
Sohka City (Takuma)	Public/Private	2 furnaces/ 150 t/d each
Fukuoka City/ Eastern Plant (Takuma)	Public/Private	2 furnaces/ 300 t/d each
Mitaka City (Takuma)	Public/Private	3 furnaces/ 65 t/d each
Ohmiya City (Kawasaki)	Public/Private	3 furnaces/ 100 t/d each
Fuji City (Kawasaki)	Public/Joint venture by public and private entities	2 furnaces/ 150 t/d each
Yokohama Hokubu (Mitsubishi/Martin)	Public/Private	3 furnaces/ 400 t/d each

Source: INFORM

MANUFACTURERS, OWNERSHIP, COSTS AND PLANT FEATURES

A large share of the waste-to-energy plants in Japan was built by five manufacturers: Takuma Co., Ltd.; Kawasaki Heavy Industries; Mitsubishi Heavy Industries; Hitachi Ship Building Co. and Nippon Kokan Steel Corp. Of the five, Takuma, Mitsubishi and Kawasaki have indigenous Japanese technology, the others being European. Takuma uses no European technology whatsoever, while Mitsubishi and Kawasaki market both Japanese and European technology. In addition, fluidized bed incineration is being developed by several Japanese manufacturers. The Machida plant is such a facility and was built by Ishikawajima-Harima Heavy Industries Co., Ltd. There are about 30 fluidized bed waste incinerators in Japan.

All incinerators in Japan are municipally owned, although one-third have contracts with private companies for operation. New waste-to-energy facilities are generally expensive to build and cost from ¥20 million to ¥30 million ($130,000 to $195,000) per design tonne, depending on amenities provided. The 300 tonnes/day Sohka (Takuma) plant, which came on line in 1985, cost ¥8 billion ($52 million). New plants include state-of-the-art combustion and pollution control equipment, such as multi-stage reciprocating grates, sophisticated air injection systems, automatic combustion controls, ash cementation or melting, acid gas controls, computer data gathering weigh-in, and truck-washing facilities.

Moreover, many facilities are not just incinerators but are community centers, often equipped with heated swimming pools and other recreational facilities. These amenities are negotiated by the community in exchange for siting a facility. INFORM visited the Fukuoka (Takuma) plant, which has a pool, tennis courts, baseball fields and a playground. The Sohka and Machida plants also have community pools. Many new facilities include classrooms, informational displays and movies about waste disposal programs in the community and the residents' important role.

All plants visited by INFORM, from the Kumamoto (Takuma) and Yokohama (Mitsubishi/Martin) plant, the two newest plants in Japan, to the 10-year-old Fukuoka plant, were in mint condition. Plant equipment was generally in top shape and many components were freshly painted. All plants looked spotless throughout and, in spite of that, were continuously being cleaned.

According to Fuji City's (Kawasaki) plant manager, Mr. Nishioka, "The city authority is very careful about how a plant looks. Nice plants are more easily accepted."

WEIGH-IN

Reflecting the Japanese attention to detailed data collection, the weigh-in station at some Japanese waste incinerators is used to collect data on the quantity of waste coming from specific locations. At the weighing platform, the truck driver inserts a computerized card into a machine and presses a button on a map to indicate the area from which the garbage has come. This enables precise data to be collected about how much garbage each area is generating. While waiting for the weight calculation, many weigh-in stations broadcast recorded messages to the truck driver, reminding him/her what can or cannot be dumped at the plant, and the penalties for violation.

THE PRIMARY PURPOSE: A WASTE TREATMENT PROCESS

Although members of Japan's government and industry often refer to the benefits of recovering energy from garbage burning, the prime purpose of incineration in Japan is to reduce the volume and toxicity of waste. After recycling, incineration is the cornerstone of Japan's solid waste management program: some materials are pre-sorted and then reused, landfilled or stored because they might have an adverse effect on incineration, and landfills are now being designed to handle large volumes of incinerator ash, which is sometimes cemented, asphalted or melted.

Commenting on why Machida City built its 2.5-megawatt, 450-tonnes/day waste incinerator, the plant manager, Mr. S. Masuda, simply stated, "pollution prevention."

And to quote from JICA:

It should be recognized that in contrast to ordinary [energy] production processes, conversion of wastes into resources has as its primary objective the adequate disposal of wastes. Therefore, even greater care should be exercised with respect to potential environmental impacts than with other ... [energy production] methods.

It would be undesirable to neglect the environmental considerations of such a process, just because a relatively higher return is attainable in terms of resource [energy] recovery.

By not operating energy production for profit and by emphasizing waste disposal from their waste to energy incinerators, the Japanese assure that the highest priority will be given to the control of pollutants, a theme repeated by each of the more than 25 workers interviewed at the eight garbage burning plants visited by INFORM.

TABLE 22 Uses of Recovered Energy

Plant	Supplies energy needs for:
Machida City	plant (including classrooms and wastewater treatment plant); greenhouse; recycling, repair and sales center; swimming pool
Kumamoto City/ Western Plant	plant (including classrooms and wastewater treatment plant); garbage truck washer; greenhouse
Sohka City	plant (including classrooms and wastewater treatment plant); heated swimming pool; garbage truck washer; sewage treatment plant and sludge drying
Fukuoka City/ Eastern Plant	plant (including classrooms and wastewater treatment plant); garbage truck washer; greenhouse; public baths
Mitaka City	plant (including classrooms and wastewater treatment plant); garbage truck washer; school and old age home
Ohmiya City	plant (including classrooms and wastewater treatment plant)
Fuji City	plant (including classrooms and wastewater treatment plant); garbage truck washer
Yokohama Hokubu	public utility, plant (including classrooms and on-site wastewater treatment plant); garbage truck washer; swimming pool; community center

Source: INFORM

Thus, while all of the 361 continuous incinerators capture and reuse energy from waste, not all generate electricity. Only 51 plants were doing so in 1982.

Another reason few incinerators generate electricity is that there is no Japanese equivalent of the Public Utility Regulatory Policies Act. The 1978 U.S. law requires utilities to purchase power from alternate energy producers at the utility's avoided cost of generation. In Japan, electric sales must be privately negotiated with utilities and utilities are not required to purchase any energy. As a result, energy is often used for other municipal purposes. For example, in Machida the incinerator provides electricity, air conditioning, heat and hot water for the plant and equipment, water treatment facility, greenhouse, workshop, classrooms, swimming pool and an arboretum (See Table 22). At other facilities energy is also provided to sewage treatment plants, homes for the aged, schools, district heating, snow melting and garbage truck washing.

OPPOSITION

Despite the priority given to controlling pollutants and despite the amenities offered, there is still widespread opposition to incineration in Japan. "Everywhere." That is how Mr. K. Nakazato, the head of overseas market development at Takuma Industries responded when asked if opposition to garbage burning plants exists in Japan. For example, the Tokyo municipal government, with only one landfill for 12 million people, has been able to site over the past 20 years only 13 of the 23 municipal solid waste incinerators it needed. Opposition exists regardless of the recycling successes that the city has achieved.

There are three reasons why incineration plants are actively opposed in Japan:

1. A history of technical defects that have caused higher pollution than necessary;

2. A concern that an incinerator might depress property values, and;

3. The nuisance of garbage truck traffic.

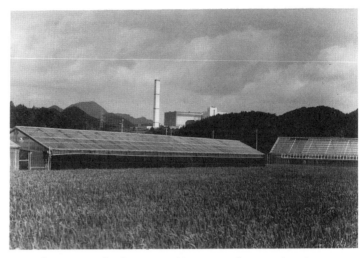

Greenhouses and other amenities — such as swimming pools and community centers — are usually offered to citizens by municipalities as inducements to accept garbage-burning plants.

Japanese garbage-burning plants are kept spotless, and visitors, such as the authors (Allen Hershkowitz and Eugene Salerni, shown here), often have to put on slippers and gloves to prevent them from getting the plants dirty.

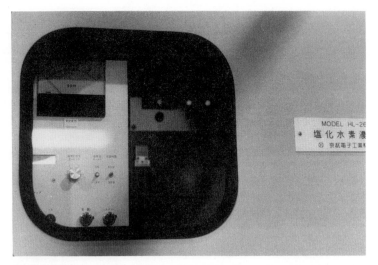

At three of the eight garbage-burning plants INFORM visited, continuous gas analyzers indicated there were non-detectable emissions of hydrochloric gas. No continuous monitoring of HCl exists at U.S. plants.

Garbage-burning plants in Japan are beautifully landscaped with arboretums and fish ponds so that they are better integrated with the surrounding community.

1. Technical Defects

Concern about an incinerator's pollution and potentially poor technological performance is, according to JICA, the "basic complaint" of citizens:

> Incineration...facilities for [rendering] waste harmless and stable...are generally considered to be a nuisance. Organized movements against them frequently surface, thus making construction difficult. The basic complaint centers around negative [environmental] impacts caused when technical defects occur...Towards this end we must ensure a preserved environment so that neighboring residents confidently understand the Agency's plans....

And past experience with waste incineration has not helped alleviate the concerns. JICA reports:

> In the past, there were many waste treatment [incineration] facilities not sufficient in function and environmental pollution control. [This] is considered to be one of the reasons citizens have sometimes had a bad image against waste treatment facilities.

2. Property Values

Residents are concerned that garbage burning plants lower property values. To counter this the local and national governments provide community amenities next to these facilities. This positive offset to the adverse affects incinerators have on property values seems to be working. According to Mr. Kato of the MHW, "in some places people actually want the plants, because of the benefits."

3. Truck Traffic

Truck traffic plagues neighborhoods where incinerators and landfills are located, but is often overlooked when communities make their solid waste management plans. This traffic not only adversely affects property values, but generates a host of environmental problems. JICA says:

> The reason for opposition to construction of incineration facilities and landfill sites is, in many cases, the traffic of collection/ transportation vehicles...Pollution caused by traffic poses a problem due to exhaust gases, noise, vibration, traffic congestion, traffic accidents, etc. Besides these, other possible impacts are effluent of filthy liquids over roads, offensive odor, etc.

Opposition to incineration is not without its practical effects. It can result in stricter emission standards, better monitoring, cleaner plants and can lead to new community amenities. Local regulations resulting from citizen pressure are invariably stricter than the national standards. According to Mr. Kato, "if municipalities do not accept the citizens' requests for standards lower than the national level, in fact, it will be impossible to site a plant. Municipalities are forced to accept the demands."

And opposition to waste incineration promotes good plant operating practices as well. Mr. Kato, who is responsible for issuing the National Government's 25% capital subsidy for waste incinerators, says that "the most important reason we monitor and keep plants clean is because of the opposition, regardless of the costs. Otherwise, we cannot build incinerators at all."

And, according to Mr. K. Onogawa, Deputy Director for water quality at the Environment Agency, "Unfortunately, we don't have a strong [unified] opposition party against the incinerator. It tends to be spontaneous, [limited] to those concerned about property values [in a certain area]. But opposition makes the plants and standards better."

ASH MANAGEMENT

Because of the widespread separation of non-combustible material from incinerator-bound waste in Japan, the volume of ash residue is less than the 10% to 20% that is characteristic of American facilities. Moreover, according to the government technical standards for new plants over 200 t/d, ignition loss, the amount of material allowed to pass through the furnace unburned, should be no more than 5% of the garbage volume introduced to the furnace (See Table 23).

TABLE 23 National Regulations for Ignition Loss

Type of Incinerator	Capacity	Ignition Loss allowed to be generated (by volume)
Continuous incineration	≥ 200 t/d	Below 5%
Continuous incineration	< 200 t/d	Below 7%
Batch type incineration	—	Below 10%

Note: Local governments may set stricter regulations than the national regulation.
Sources: Environment Agency and Clean Japan Center

Incinerator ash is disposed of in a variety of ways by the Japanese. Fly ash is often mixed with bottom ash (considered less hazardous than fly ash) and disposed of together in a landfill that controls leachate. Nevertheless, there is some concern over the leaching of toxic substances from ash, and the cementing, asphalting and melting of ash is gaining popularity. Ash cementing or asphalting is not required nationally, although many municipalities demand this precautionary measure be carried out. When asked why it is done if it is not required, Mr. K. Hiroshima of Kawasaki Heavy Industries, said, "People are concerned about heavy metals leaching, so we do it. And also, with briquetting, the land[fill] can be reclaimed more efficiently."

Several different methods of ash disposal were practiced at the eight plants visited by INFORM (See Table 24).

At the Fukuoka and Kumamoto (Takuma) plants, fly ash and bottom ash are mixed, given moisture conditioning to reduce dust and then sent to a nearby landfill equipped with liners and leachate collection and treatment systems. An interesting feature of the Fukuoka plant and all new Takuma plants is that grate siftings, often unburned, are returned by conveyor to the refuse pit where they will eventually be reintroduced to the incinerator. This technique keeps the ignition loss (volume of ash, or unburned waste) at this plant below 2%, a very low value.

At the Machida (IHI), Mitaka (Takuma) and Ohmiya (Kawasaki) plants, fly ash is injected with 10% to 20% cement and water. It is then subjected to pressure, hardened into pellets or bricks, mixed with bottom ash and sent to a lined landfill with leachate collection and wastewater treatment systems. At the Mitaka plant, this process adds ¥48 (31 cents) per tonne of refuse processed. The Japanese believe that this produces an environmentally stabilized product.

The Sohka (Takuma) plant is equipped with ash melting equipment, that processes incinerator bottom ash into a granulated glass slag to prevent the leaching of heavy metals at the landfill. Ash is reduced to one-third of its original volume and becomes only a small percentage of the volume of the original waste. The ash remelting furnace is oil fired and operates at 1250°C to 1350°C (2300°F to 2500°F). Exhaust gases are routed back through pollution control equipment — electrostatic precipitator and wet scrubber — for treatment. The process costs ¥20,000 ($130) per tonne of ash, or $20 to $30 per tonne of refuse.

**TABLE 24 Pollution Control Devices
and Ash Handling Practices at Plants Visited by INFORM**

Plant	Pollution control	Ash handling
Machida City	3 field ESP; Calcium oxide (lime) injection into fluidized bed furnace at 20-25 kg per tonne of refuse for acid gas control; on-site wastewater treatment facility; noise muffler; 59.5 meter stack*	Fly ash (90%) gets mixed with cement (10%) to reduce leaching potential and is pelletized, mixed with untreated bottom ash and taken in covered trucks to class "C" domestic waste landfill with liner and leachate control
Kumamoto City/Western Plant	2 field ESP; wet scrubber; on-site waste-water treatment facility; sump pump in garbage pit to drain leachate and inject it into furnace; noise muffler; 59.5 meter stack*	Fly ash and bottom ash mixed in no specific ratio and taken in covered trucks to class "C" landfill with liners and leachate control
Sohka City	3 field ESP; wet scrubber; on-site wastewater treatment facility; sump pump in garbage pit to drain leachate and inject it into furnace; noise mufflers; 59.5 meter stack*	Bottom ash melting at 2300 - 2500°F (1250-1350°C) to reduce ash volume by 2/3 and to reduce the possibility of leaching; fly ash taken in covered trucks to class "A" hazardous waste landfill
Fukuoka City/ Eastern Plant	3 field ESP; wet scrubber; on-site wastewater treatment facility; sump pump in garbage pit to	Fly ash and bottom ash are mixed and moisture conditioned to reduce dust in no specific ratio and taken in covered trucks

This conserves valuable landfill space, reduces disposal costs and controls leaching.

While the ash melting furnace at Sohka can receive both bottom and fly ash, fly ash is not processed because it is more difficult to melt and the dust was causing problems. However, according to Mr. K. Onogawa, heavy metals leaching tests of melted fly ash showed the ash met national leaching standards.

Other plants do melt fly ash as well as bottom ash but local regulations and economics determine the arrangement. Officials

Table 24 continued

Plant	Pollution control	Ash handling
	drain leachate; noise muffler; 80 meter stack	to class "C" landfill with liners and leachate control
Mitaka City	3 field ESP; wet scrubber; on-site wastewater treatment facility; sump pump in garbage pit to drain leachate and injects it into furnace; noise mufflers; 59.5 meter stack*	Fly ash (80-90%) gets mixed with cement (10-20%) to reduce leaching potential and is made into bricks before being mixed with untreated bottom ash and taken in covered trucks to class "C" waste landfill with liners and leachate control
Ohmiya City	2 field ESP; dry scrubber; on-site wastewater treatment facility; noise muffler; 55 meter stack*	Fly ash (80%) is mixed with cement (20%) to reduce leaching potential and is made into bricks before being mixed with untreated bottom ash and taken in covered trucks to class "C" waste landfill with liners and leachate control
Fuji City	2 field ESP; dry scrubber; on-site wastewater treatment plant; noise muffler; 59.5 meter stack*	Fly ash and bottom ash are mixed in no specific ratio and taken in covered trucks to class "C" landfill with liners and leachate control
Yokohama Hokubu	3 field ESP and cyclone collector; semi-dry scrubber; on-site waste water treatment plant; noise muffler; 130 meter stack; de-NO_x system	Fly ash and bottom ash are mixed in no specific ratio and taken in covered trucks to class "C" landfill with liners and leachate control and wastewater treatment

* Because all smokestacks above 60 meters are required to be painted red and white for purposes of aviation protection, it is common practice for waste incinerators to employ 59.5 meter stacks.

at the Sohka plant tried to sell the vitrified bottom ash for alternative uses but have found it "politically difficult to do so" because of concern about contamination. Ash is melted at eight facilities in Japan and the technique is gaining in popularity. Takuma built three of these facilities, Nippon built two, and Showadenko, Tsukishima and Kubota built one each. They are installed at incinerators ranging in capacity from 20 to 450 tonnes/day. Another is under consideration for the 1,800 tonnes/day Tokyo facility. A variety of fuels is used to remelt the ash, including heavy oil, kerosene, added oxygen (PSA process) and, under development, electricity. Takuma markets a unit which can be retrofitted. According to Mr. Nakazato of Takuma Industries, uses such as road base material have been found for melted ash from other plants, although, according to Mr. Onogawa, this is not widely practiced.

TELEMETERING AND MONITORING

All plants in and around Tokyo and some in other parts of Japan, including the Mitaka plant visited by INFORM, have telemetering systems so that they can be in continuous communication with the prefectural environment agency on air quality. If ambient air emissions deteriorate, the plant can be instructed to take emissions reduction measures based on four grades of air quality alert:

1. Caution. Informs the plant operators of adverse ambient air conditions but requires no reduction in plant operations or emissions.

2. Alarm. Requires the plant to reduce throughput to the furnace by 20%.

3. Emergency. Requires the plant to reduce throughput to the furnace by 30% in the summer and by 50% in the winter when more heating furnaces are in use.

4. Shutdown. (According to Mr. H. Ogasawara: "When an incinerator is shut down unexpectedly, the collected household wastes are incinerated at the closest plants. This kind of agreement is made between local governments or plant operators.")

Continuous emissions monitoring indicates not only the level of pollutants going into the atmosphere, but also helps determine how well the control equipment and the furnace are working. All of the eight plants visited by INFORM monitor hydrogen chloride, oxides of nitrogen and sulfur, and temperature. At three of the plants the hydrogen chloride monitors indicated non-detectable emissions, with detectability being 1 ppm. All, with the exception of Fukuoka, monitor oxygen (Fukuoka monitors carbon dioxide, as does Mitaka) and all have full color television monitors continuously displaying the inside of the furnace and the stack exit (See Table 25).

TABLE 25 Continuous Monitoring of Pollutants and Combustion at Plants Visited by INFORM

Machida City	Temperature (calculated by 5 sensors in each furnace); O_2; NO_x; SO_x; HCl; stack opacity; ESP pressure; color TV screens continuously display inside of furnace and stack exit
Kumamoto City/ Western Plant	Temperature; O_2; NO_x*; SO_x*; HCl**; color TV continuously display inside of furnace and stack exit; automatic combustion control***
Sohka City	Temperature (2 areas of furnace); O_2; NO_x; SO_x; HCl**; Ph monitor for scrubber; color TV screens continuously display inside of furnace and stack exit; automatic combustion control***
Fukuoka City/ Eastern Plant	Temperature (3 areas of furnace); CO_2; NO_x; SO_x; HCl; color TV screens continuously display inside of furnace and stack exit
Mitaka City	Temperature (2 areas of furnace); CO_2; O_2; NO_x; SO_x; HCl**, color TV screens continuously display inside of furnace and stack exit
Ohmiya City	Temperature (2 areas of furnace); O_2; NO_x; SO_x; HCl; color TV screens continuously display inside of furnace and stack exit
Fuji City	Temperature (2 areas of furnace); O_2; NO_x; SO_x; HCl; color TV screens continuously display inside of furnace and stack exit
Yokohama Hokubu	Temperature, O_2; HCl; SO_x; NO_x; color TV screens continuously display inside of furnace and stack exit.

* Continuously displayed outside of control room because plant management states that these emissions are too low to be significant, usually being emitted at levels that are well below regulatory requirements.

** Monitor indicated "non-detectable" emissions of HCl at time of visit.

*** ACC is employed to maintain a constant steam temperature and pressure despite the temperature fluctuations that inevitably occur when burning a heterogenous fuel such as garbage. It does so by adjusting the garbage feed rate based on data provided by a "finishing-line sensor." Through the use of "burn out" weight measurements the "finishing-line sensor" calculates the average caloric value of the garbage being burned and then adjusts the garbage feed rate.

DIOXIN EMISSIONS

Dioxins and furans were first found in the emissions of municipal waste incinerators in 1977, but were not detected in Japan until 1983. Media reports on this raised public concern, and an "Experts' Committee" was established in the Ministry of Health and Welfare, which released its findings in early 1984. The report analyzed dioxin in fly ash and, through a number of assumptions, estimated the human dosage to residents living in the vicinity of a plant. This was found to average less than the Swedish government's recommended limit of 0.1 nanograms of 2,3,7,8-TCDD per kilogram of body weight per day. The committee concluded that no adverse health effect existed from these incinerator emissions. A research study launched to measure actual flue gas concentrations at 33 municipal waste incinerators concluded that "TCDD doesn't have a negative influence on human beings' health...but it is necessary to continue to study dioxin's influence." The Environment Agency also tested four incinerators and found comparable emissions levels.

In these studies, specific plants were not identified and only ranges of test results were presented. Table 26 compares the Japanese findings with plants in New York equipped only with electrostatic precipitators, and a facility in West Germany equipped with an acid gas scrubber and baghouse and found to be emitting extremely low levels of dioxins. The Japanese levels span the range from very low to very high. This is what is to be expected since the Japanese tests included a representative cross section of incinerators, from older types with unsophisticated pollution control devices to modern facilities equipped with acid gas scrubbers and efficient particulate removal systems.

However, critics of Japan's dioxin research, including some of the members of the Swedish Environmental Protection Board, have expressed concerns about Japanese sampling, analytic and risk assessment methods. This, combined with the lack of reported details (such as which plant was tested, its equipment type, and combustion conditions) calls into question the usefulness and comparability of these test results to plants outside Japan.

In a recent follow-up study, two incinerators that were tested in the 1984 MHW study were retested and the samples analyzed using a different method (capillary column method). MHW researchers found lower levels of dioxin and "considered that these differences ... clearly depend on the analytical method."

TABLE 26 Dioxin Flue Gas Concentrations at Japanese Municipal Waste Incineration Facilities

Contaminant	Japanese Test Results*		Comparative International Test Results		
	1984[1] Range(ng/Nm³)	1986[2] Range(ng/Nm³)	Albany ANSWERS[3] Av.(ng/Nm³)	Westchester RESCO[4] Av. (ng/Nm³)	Wurzburg West Germany[5] Av.(ng/Nm³)
2,3,7,8-TCDD	N.D. (1 sample)	0.28 - 1.20 (3 samples)	0.38	0.28	0.018
Total TCDDs	N.D. - 109 (25 samples)	0.66 - 175 (5 samples)	16.6	2.85	1.91
PCDDs	133 - 13,600	101 - 924	509.6	24.02	22.10
PCDFs	436 - 10,000	231 - 762	92.2	76.23	27.85
Eadon Toxic Equivalents[6]	NA	NA	16.0	3.83	0.81

1. A study of 33 municipal refuse incinerators for the Japanese Ministry of Health and Welfare.
2. A study of 4 incinerators for the Japanese Environment Agency.
3. The ANSWERS plant is equipped with ESPs. Average of 3 tests, auxiliary gas on.
4. The Westchester plant is equipped with ESPs and has the lowest dioxin emissions reported in New York State. Average of 3 tests.
5. The Wurzburg plant is equipped with an acid gas scrubber and baghouse.
6. The toxicity of a mixture of PCDDs and PCDFs as compared to the toxicity of the 2,3,7,8-TCDD isomer, according to the NYS Department of Health method.
N.D. – not detectable
NA – not available
ng/Nm³ – nanograms per normal cubic meter.
* Sampling and analytical techniques may differ from those used in other countries.
Compiled by INFORM

Detailed data were also reported, but while it may seem that the Japanese are now entering the mainstream of dioxin research, the authors concluded, "The accurate data of dioxins from ... [Japanese] municipal incinerators are quite ... few compared with the European countries and the U.S. We need ... more data and ... more research on this problem."

In 1986, the Japanese hosted the 6th International Dioxin Conference in Fukuoka and two additional dioxin conferences in Machida and Kyoto, all attended by INFORM. And they plan to continue studying this issue. In spite of Japan's history of and high reliance on incineration, especially in urban areas, it is important to note that, as pointed out by Dr. Christoffer Rappe of Sweden at the 6th International Dioxin Conference, the levels of dioxin found in human tissue in Japan are similar to those found in other industrialized countries. Also, dioxin concentrations found in urban residents are higher than in rural populations.

WORKER TRAINING

All workers at incineration plants in Japan must undergo special training, sometimes referred to as "safety supervisory" training. According to Mr. K. Onogawa of the Environment Agency, solid waste workers "have to go through three feet of textbooks" relating to combustion, pollution control and safety. Training courses at garbage burning plants are provided by the Japan Environmental Sanitary Association.

But according to Mr. K. Hiroshima, who used to head Kawasaki's Overseas Marketing Group and guided INFORM through two of his company's waste-to-energy plants, "Worker training is very difficult. It is a problem because there are many sub-systems at waste-to-energy plants that may not be connected to each other. For example, we must install wastewater treatment plants, and we must explain this to the workers. But it is completely different from burning garbage."

The complexities involved in operating a garbage burning plant cause fully one-third of all plants in Japan, which are all municipally owned, to contract out the day-to-day operation to private firms. Mr. K. Niskioka, who oversees the operation of the Fuji City plant for the city, explains why operation was given to a private company. "These systems are very difficult to run safely and worker education is a big problem for the city. So we

Assigning a large staff to clean apparently clean windows
and floors in a plant is a common practice in Japan.

A truck-weighing station at a plant is also beautifully
landscaped.

Japanese garbage-burning plants are spotless.

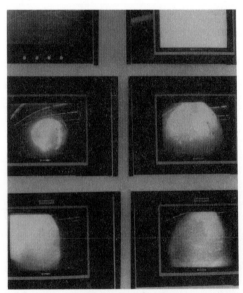

Full color monitors show furnace combustion to help plant operators reduce pollution.

contracted a subsidiary of Kawasaki [which built the plant]."

There are three reasons for training of workers in solid waste management. First, special training makes it more likely that workers will do their job correctly and conscientiously, thereby reducing threats to the public health from the processing and disposal of municipal waste. Second, well trained workers can better explain waste disposal to the citizenry and so help generate public cooperation. And third, familiarization with the hazards inherent in the work can prevent illness or injury.

Five laws cover training requirements for workers at incineration facilities. They include the Labor Safety and Sanitation Law, Waste Disposal and Public Cleansing Law and Enforcement Regulation of the Law. The following work classifications at garbage burning plants are subject to specialized training regulations:

Safety specialists

Sanitary specialists

Fire specialists

Dangerous-object handling technicians

Boiler technicians

Crane operators

Electrical engineers

Turbine engineers

However, according to Mr. Kato of the MHW, "many workers who are not required to take certain courses do so anyway."

Protecting Public Health and the Environment

Workers with control room responsibilities at incinerators are required to have a college degree in engineering and six months additional classroom instruction unless they have previous garbage-burning experience. With experience, only two weeks instruction is required. Instruction is designed to familiarize plant operators with burning a heterogenous substance such as garbage, how to control the hazardous emissions and how to handle and dispose of the dangerous ash residue.

Landfill operators also receive special certification that requires education and training to prevent public health threats caused by mixing reactive wastes. According to JICA:

Generally speaking, reactive waste accidents ... [that] generate ... poisonous gases ... occur much more often because waste management workers are poorly [trained or] supervised than [because] waste generators" improperly mix wastes.

Achieving Citizen Cooperation

Workers are also trained to educate the local population about waste disposal, recycling, landfilling and incineration techniques. At each of the eight incinerators visited by INFORM, classrooms with advanced audio/visual equipment were used by plant operators to explain their facilities. They can also explain waste management in their city with an advanced understanding of all the issues, including monitoring, emissions control, pre-sorting and ash handling. The Japanese view workers involved in solid waste management as ambassadors who are often in a good position to help generate local cooperation in an often highly coordinated solid waste management strategy. A JICA publication says recycling, landfill or incineration operators:

can be a prominent representative of a city. Appearance, courtesy, competence and consideration on the part of the...worker provide the first and foremost incentive (or disincentive) to citizen cooperation with the refuse system. Therefore the initial step in upgrading the system so that it is a mutually beneficial cooperative effort between local government and urban residents is proper training and supervision of workers.

Reducing Worker Accidents

Waste disposal is a dangerous occupation. JICA reports:

Workers engaged in refuse collection, recycling or ... disposal ... are always exposing themselves to ... harmful and/or toxic wastes directly or indirectly ... Municipal refuse may contain dangerous materials such as broken glass, metal, used syringes or razors which may cause injury to workers. Municipal refuse also may contain used paper diapers, and hospital or clinical waste, (such as medical appliances, radioactive materials and sometimes infectious substances). Pathogenic bacteria contained in such wastes is likely to migrate in the air and can infect workers ... [Although] surveys show that accidents caused by hazardous material ... are much more due to 'mixed collection' than 'separate collection' ... workers ... handling wastes directly at a [recycling] facility where bottles, glass, plastic, cartons and metal are recovered from waste are [also] in danger of injury or illness ...

Incinerator operators are trained, or workers with previous experience with burning garbage are sought, "to control environmental pollution and ensure labor safety" (JICA). And workers who operate landfills must be trained because ammonia, hydrogen sulfide and explosive methane are produced at landfills and, according to JICA, "workers' health care should be considered as they are constantly exposed to such gases [and they] ... are in danger of being poisoned by these substances."

Despite Japan's widespread training of solid waste management workers, the accident rate for workers in public sanitation services, including collection, transportation, processing and disposal, is greater than the rate in any other Japanese industry (See Table 27). However, about 75% of the accidents in solid waste management occur during the collection and transportation of wastes, while only about 10% are associated with the incineration process. JICA quotes the three most common reasons why worker accidents occur:

1. Collection and transportation of waste [including recycling] has to be done on public roads

2. There are some refuse pits ... ash pits ... or waste water tanks [at incinerators or landfills] which may generate hydrogen sulfide or cause oxygen deficiency, and in addition [workers must contend with] high temperature incinerator equipment and high altitude work places

3. Some refuse contains hazardous substances or heavy materials which cause illness or injury.

TABLE 27 Number of Labor Accidents in Japan per 1,000 Employees per Year

Sanitation	63.3
Policemen	27
Firemen	19
Seamen	16.6
Utilities (gas, water, electricity)	11.7
Transportation	11.3

Source: JICA. According to the JICA "New York City has similar ratios, though the data for that city are not up-to-date."

JICA says physical injury to solid waste management workers is "the highest-occurring ailment... [that] can be prevented to some extent by performing ... precise health examinations ... and by rearranging the age structure of workers. Calisthenics before working is necessary for all forms of physical work."

Although JICA has said "it goes without saying that the causes of accidents are due to bad working environments," the agency has specified two categories of causes of accidents in solid waste disposal: human factors and physical factors. Human factors include carelessness, lack of concentration, and inappropriate motion. These causes are avoidable and, in part, form the basis of worker training programs. The physical factors contributing to accidents "include original [technological] defects of... the facilities, or [the] lack of necessary safety systems."

REGULATIONS FOR MUNICIPAL SOLID WASTE INCINERATORS

National Regulations

The Basic Law for Environmental Pollution Control enacted in 1967 established the first national emission standards specifically for municipal solid waste incinerators in Japan. According to paragraph 1, article 9 "the Government shall establish environmental quality standards....relating to air, water and soil pollution and noise ... the maintenance of which is desirable for the protection of human health and the conservation of the living environment." The law states that whenever new scientific knowledge is obtained, these "standards shall be revised whenever necessary". Since 1962, sulfur dioxide regulation in Japan has been made more stringent on eight occasions, and since 1973 nitrogen oxide regulation has been made more stringent on five occasions. As SO_x, HCl and particulates standards evolved, but not in the case of NO_x control, existing energy-recovering incinerators had to retrofit their technology to upgrade their performance.

The 1967 law controlled emissions from municipal solid waste incinerators on the basis of individual sources of pollutants. However, this was changed in 1974 to take account of ambient air quality. According to a January 1985 document prepared by the Air Quality Bureau of the Environment Agency, the "individual-source" approach:

proved to be ... ineffective because in areas where a large number of pollutant sources were concentrated, they added up to a substantial amount, however insignificant the emission quantity of individual sources. [Moreover], where individual sources grew rapidly in scale, [overall] emission[s] ... increased in spite of individual source control. To remedy such shortcomings of the individual [source] approach, the concept of an overall [ambient air] approach ... was incorporated into the Basic Law in 1974.

Regional and Municipal Regulations

In the U.S., Federal Clean Air Act and Clean Water Act regulations can be made more stringent by the states or municipalities. A similar regulatory arrangement exists in Japan, and that country's nationwide environmental regulations are invariably supplemented by more stringent environmental regulations issued and enforced by the 47 prefectural governments and 3,255 municipalities.

National emission standards for municipal solid waste incinerators are with few, if any, exceptions exceeded by stiffer standards set by prefectural governors and, in most cases, even stiffer standards set at the municipal level. For example, although Japan's nationwide HCl limit is 430 ppm, none of the eight municipal incinerators visited by INFORM exceeded an HCl limit of 100 ppm, and most were in the 25-50 ppm range. (The emissions limits of four of these plants are shown in Table 28.) Mr. K. Nakazato of Takuma Industries suggests that in the absence of health effect data, municipal regulations are the most stringent

TABLE 28 Emissions Limits of Four Selected Japanese Waste Incinerators

Facility	Particulates (g/Nm^3)	$(gr/dscf)$	HCl (ppm)	NO_x (ppm)	SO_x (ppm)
Machida City	0.02	(0.01)	NA	NA	NA
Fukuoka City/ Eastern Plant	0.03	(0.01)	30	NA	NA
Sohka City	0.05	(0.02)	25	200	30
Mitaka City Plant	0.03	(0.01)	25	120	30

NA — not available
g/Nm^3 — grams/normal cubic meter
$gr/dscf$ — grains/dry standard cubic foot
ppm — parts per million
Source: INFORM

Regional Regulations

A 1974 amendment to the Basic Law for Environmental Pollution Control required that sulfur oxide emissions from municipal solid waste incinerators be established on the basis of "area regulation." The same system was applied in 1981 to emissions of nitrogen oxides. There are 24 regulated regions in Japan for SO_x and three area regulated regions for NO_x.

An "area" is a jurisdiction comprising territory from within a prefecture, and sometimes from more than one prefecture, established by the Japan Environmental Agency for the specific regulatory purpose of controlling SO_x and NO_x. Applicable regions, according to the Environmental Agency are "those [areas] where plants and businesses are clustered and it is recognized that the ambient air quality standards will not be attained by application of existing regulations alone." Area regulation requires prefectural governors, after considering health effects and technological feasibility, to establish SO_x and NO_x regulations for each of their municipal solid waste incinerators by establishing an area SO_x and NO_x limit in consultation with other governors or mayors in the area.

Area regulation was established because in some political jurisdictions the emissions standards were not sufficient to ensure air quality standards set by the national government. For facilities within an area, allowable SO_x and NO_x emissions are computed by assessing total emission sources as well as the area's meteorological and topographic features.

because "the politics are more intense." And according to JICA:

Except in the case of sulfur oxides, the prefectural government is authorized to establish more stringent emission standards instead of the national emission standards ... [when] national ... standards are recognized to be insufficient in the specific area.

Although the national regulations for municipal solid waste incinerators include only HCl, particulates, SO_x, NO_x, and wastewater (see Table 29, page 97), the latitude that prefectures enjoy in making more stringent regulations has resulted in some incinerators being regulated for heavy metals as well. According to Mr. Kato, of the Ministry of Health and Welfare, "prefectural governments have their own, more strict regulations and so in some places more pollutants are controlled, for example, metals

Photo: Kawasaki Industries

Citizens' concerns about permit violations are now alleviated by continuous outdoor public displays of emissions data recorded by garbage burning plants, such as the data shown on this billboard outside Kawaski's plant in Kyoto.

Beautifully designed interiors of garbage-burning plants encourage citizens to accept these facilities.

Workers in typical control rooms in Japanese garbage-burning plants pay meticulous attention to monitors at all times.

Refuse pits are enclosed by steel doors to keep tipping bays safer and cleaner.

such as cadmium, lead and zinc." However, although each of the 47 prefectures is free to set its own more stringent emissions regulations, only minor regulatory differences exist in substance among the prefectures. And this is so whether the prefectures are urban or rural. Mr. Kato says,

> Although Tokyo prefecture has the most stringent [prefectural] regulations the differences between the most stringent [prefectural] regulations and the least stringent is not so much because even in less populated areas the citizens demand the best control possible. Even people in the countryside demand that their local plants do what is done in Tokyo. Otherwise, legal actions are taken.

Basis for Municipal Standards

Municipal governments can implement pollution control regulations more stringent than the national government or the prefectural governments, because they can best take "into account the specific natural and social conditions of the area concerned," says JICA. Because municipalities take "social conditions" into account, local regulations, unlike national regulations, tend not to be based on correlations with health effect data. Instead, they are more often based on what the local population demands in exchange for allowing an incinerator to be sited, and on how well a specific technology is known to perform. This usually results in emissions much lower than the levels that the Japanese government substantiates as being safe. Commenting on the regulatory standards on the local level, Mr. Nakazato of Takuma Industries said "It just makes political sense."

National Government Demands That Its Own Regulatory Standards be Surpassed

Although prefectural and municipal governments have authority to enforce the most stringent regulations for solid waste incinerators, in practice the national government itself demands that incinerators meet standards more severe than those it has officially established. This interesting enforcement of regulations is due to three factors. First, Article 24 of the Basic Law for Environmental Pollution Control states that the government "shall endeavor to take the necessary measures to encourage the ... improvement ... of facilities for the prevention of environmental pollution." This encourages the national government to subsidize technologies that can exceed those regulations that health effect data can substantiate. Second, alleviating the widespread, though usually spontaneous and unorganized, opposition to waste incineration demands that the lowest achievable emissions rate be aimed for. Third, the government is interested in advancing the performance of emissions control technology and solid waste incinerators so that the technology might become more internationally competitive and exportable.

Subsidies

The Waste Disposal and Public Cleansing Law of 1976 requires "the national government ... to aid municipalities financially [for] a portion of [the] expenses for construction of solid waste disposal facilities." Currently, the national government subsidizes 25% (up to 50% in polluted areas) of the capital costs involved in plant construction and guarantees the municipal loans that cover the rest. The Ministry of Health and Welfare administers the national subsidy and loan program and the Head of the Waste Management Section approves the subsidies. Mr. Kato, the former head, says that because "the national government is in the position to subsidize local plants, and usually the plants cannot be built without our help ... opponents come [to Mr. Kato] and ask that I not give the subsidy [or loan guarantee] unless the plan will guarantee the lowest emissions. And I agree, the technology can do it and it can always do better. It is in our interest to see it do better."

According to Mr. Kato, although national standards for emissions from municipal solid waste incinerators are established to conform with "safe health levels," regulatory standards at the prefectural and municipal level are established to address citizens' concerns which would otherwise prevent the siting of these plants. Because he believes the technology can do it, Mr. Kato thinks that citizen demands for stricter regulations can be met.

Nitrogen Oxides Regulations

NO_x emission standards were first established in Japan in 1973 and were not applicable to waste incinerators until 1977. NO_x emissions, in particular from automobiles, are Japan's gravest air pollution concern, although leaded gasoline is still the predominant automotive fuel No NO_x reduction technologies of any sort are required on automobiles.

Since 1973 Japan's NO_x standards have been upgraded on five occasions, but not because of adverse health effects. Instead, the NO_x standards were upgraded for technology-based issues. According to the Environment Agency's Outline of Air Pollution Control in Japan:

> The latest revision [5th upgrading, September 1983] was made in response to (1) changes in energy supply from oil to solid fuels [coal and solid wastes] which generate more nitrogen oxides, and (2) technical progress in combustion technology to reduce NO_x emissions.

NO_x is the only pollutant for which the upgrading of regulatory standards did not require the retrofitting of existing plants.

Japan's NO_x emission standards for municipal solid waste incinerators vary with the age of the plant and its size as calculated by flue gas emissions rate (See Table 29). If an incinerator

TABLE 29 National Emission Standards for Municipal Solid Waste Incinerators

Category	Flue gas volume*	Emission Standard**
Particulates		
Ordinary	$\geq$40,000 Nm³/h	0.15 g/Nm³
Standards:	<40,000 Nm³/h	0.50 g/Nm³
Special	$\geq$40,000 Nm³/h	0.08 g/Nm³
Standards:	<40,000 Nm³/h	0.15 g/Nm³
NO_x (at 12% O_2) Plant Constructed:		
On or before June 18, 1977	$\geq$40,000 Nm³/h	300 ppm
After June 18, 1977	$\geq$40,000 Nm³/h	250 ppm
On or before August 10, 1979	<40,000 Nm³/h	300 ppm
After August 10, 1979	All plants	250 ppm
HCl		
For all facilities in all areas:		700 mg/Nm³ (430 ppm)
SO_x		
Varies according to "K-value," which takes account of the region in which the plant is located, stack height and the velocity of emission gases.		

* A flue gas volume of 40,000 normal cubic meters/hour is emitted by plants burning about 10 U.S. tons an hour or 250 U.S. tons of garbage each day.

** Unit is grams/normal cubic meter. A gram/normal cubic meter is equivalent to 2.3 grains/dry standard cubic foot.

Source: Environment Agency

TABLE 30 Desulfurization and Denitrification at Municipal Solid Waste Incinerators in Japan (1986)

	Number of installed units	Total treatment capacity $(10^3 Nm^3/h)$***	Total treatment capacity (tonnes/day)	Average removal efficiency
SO_x*	52	2,996	57,600	73.7%
NO_x **	4	210	5,200	54.3%

*   Includes wet and dry scrubbers. Represents 3.8 % of all desulfurization units employed by all Japanese industry.

**   Includes only ammonia catalytic reduction process. Represents 2.1 % of all denitrification units employed by all Japanese industry.

*** 1,000 normal cubic meters/hour

Source: JICA

with a stack gas emissions volume of greater than 40,000 Nm^3/h (equivalent to burning about 10 U.S. tons/hour) was built before June 1977, its national NO_x standard is 300 ppm. If such an incinerator was built after June 1977, its NO_x standard is 250 ppm. For plants emitting less than 40,000 Nm^3/h built before August 1979, the NO_x emissions standard is 300 ppm. And for all plants built after 1979 regardless of their size, the NO_x emissions rate is 250 ppm.

However, a NO_x emission rate of 250-300 ppm is usually the level emitted by incinerators not using any de-nitrification equipment, so the Japanese standard involves only a very small, if any, percentage reduction in actual NO_x emissions from incinerators. Nevertheless, some officials in Japan's garbage burning industry are not concerned about even stricter regulations. Commenting on Tokyo's and Osaka's NO_x limit of about 70 ppm, which is one third the national standard, Mr. K. Nakazato of Takuma said, "This is not so difficult to achieve." He points out that because "Tokyo wants NO_x emissions guaranteed at 70 ppm, others will work to develop the technology." Only four recently built incinerators in Japan actually use de-nitrification equipment (See Table 30).

Sulfur Oxides Regulations

Emission standards for SO_x were first introduced in 1972 and have been upgraded eight times since then. The regulations vary with the location, height of smokestack and stack gas emission rate. The Japanese use a formula known as the "K-value standard" to calculate the amount of sulphur oxides per cubic meter of gas per hour that may be emitted from a municipal solid waste incinerator.

The formula is:

$$Q = K \times 10^{-3} \times He^2$$

Q = allowed emissions volume of SO_x per normal cubic meter per hour (Nm^3/h).

K = a constant established by the Environment Agency for the area in which the incinerator is located.

He = stack height in meters which is adjusted to take into account the plume rise and velocity of emission gases in meters per second.

There are 16 categories of "K-values" with numbers that range from 3.0 to 17.5. In areas where SO_x levels are exceptionally high, such as Tokyo, Osaka, Nagoya and Yokahama, three classes of "special K-value" standards further limit SO_x emissions beyond the standard K-value regulation, sometimes by as much as 50%. These special standards are used, sometimes for only a limited time, in case a prefectural governor recognizes that "seasonal fluctuations" or facility emissions will adversely affect ambient air quality. If a prefecture's air quality is in chronic non-compliance with ambient air standards, the governor may require the facility to modify its fuel content to achieve compliance or shut down. However, no municipal solid waste incinerator has ever been closed because of emissions violations. There are fines for violating the SO_x standard. The proceeds of these fines go to a fund used to assist the 90,000 certified victims of environmental pollution.

K-value emission standards for SO_x apply regardless of the type of fuel used. Fifty two of Japan's 361 energy-recovering municipal solid waste incinerators use acid gas scrubbers (See Table 30).

Particulate Regulations

Japan maintains two national emission standards for particulates. Like all emission regulations, these are invariably surpassed by more stringent standards on the prefectural and municipal level. The most widespread standard, Japan's "ordinary emission standards," apply uniformly throughout the

Table 31 Types of Particulate Control Equipment

Extent of installation of particulate collecting equipment at 359* municipal solid waste incinerators in Japan with a capacity of 40,000 Nm^3/h (250 U.S. Tons/day) or more.

Type of Equipment	Percent of Incinerators Equipped
ESP	71.6 %
Wet collection device	14.0 %
Centrifugal collector	11.8 %
Filtrating collector	1.3%
Others	1.3 %

* Two recently [1986] built facilities are not included here. These facilities, visited by INFORM, each employ an ESP.
Source: JICA

country. For incinerators that burn 250 U.S. tons/day or more, the ordinary particulate standard is 0.15g/Nm^3. For incinerators burning less than 250 U.S. tons/day the particulate standard is 0.50g/Nm^3.

The second set of regulations are called "special emission standards." These apply to facilities built after May 1982 and located in 12 areas where air pollution is most severe. For incinerators burning more than 250 U.S. tons/day the standard is 0.08g/Nm^3 and for those that burn less than 250 U.S. tons/day the special emissions standard is 0.15g/Nm^3 (See Table 29). Japan has different particulate standards for furnaces in other industries, and the particulate standard for waste incinerators is the least stringent of all. The most commonly used particulate control device is the ESP (See Table 31).

Hydrogen Chloride Regulations

In addition to regulating NO_x, SO_x and particulates, the Air Pollution Control Law designates four groups of emissions as toxic substances requiring control. However, only one group, HCl, is applied to waste incinerators, and the national emissions standard is 700 mg/Nm^3 or 430 ppm (See Table 29). HCl is regulated even though Mr. Kato of the Ministry of Health and Welfare has stated that "Japan doesn't have an acid rain problem as [the U.S. does]. We don't see acid rain damage even though 68% of our land is covered by forest." When the HCl regulation went into effect in 1978, all of Japan's incinerators were required to take action to reduce HCl emissions to meet the new standard. According to Mr. Kato:

When the HCl regulations went into effect the older plants had to retro-fit or otherwise take action to reduce HCl emissions. Today, all Japanese plants are either equipped with HCl control equipment or separate plastics to reduce HCl emissions. And this is true even for small plants, [that burn] below 50 tonnes per day.

EMISSIONS TESTING

Emission gases from garbage burning plants are tested frequently. According to Mr. T. Hayase, who heads the Air Resources Division of Japan's Environment Agency, emission gases must be analyzed at least once every two months for large facilities (those with a gas emissions volume of 40,000 Nm^3/h or more) and once every six months for small facilities. These periodic analyses must include tests for particulates, NO_x, SO_x and HCl.

VII

LANDFILLS

Due to the extent of recycling and incineration, only about 10%-20% of the country's municipal solid waste is sent directly to landfills unprocessed (compared to more than 90% in the U.S.) The landfilled material consists mostly of incombustibles and incinerator ash. In 1983 there were 2,679 municipal waste landfill sites with a total capacity of 383 million cubic meters (See Table 32). These sites had space for another 171 million cubic meters in 1983, which officials estimated would be used up in six to seven years. It is now extremely difficult to acquire new landfill sites.

TYPES OF LANDFILLS

Japan has three basic landfill types, each with its own applicable standards:

Type A

Toxic waste landfills provide concrete box separation of toxic and hazardous materials (See Charts 9a, 9b, 9c).

Type B

Stable waste landfills receive nunputrescible waste such as glass, plastics, and construction and demolition debris.

Type C

Leachate control landfills receive raw waste, incinerator ash, and other nonhazardous waste. These facilities are required to have liners and leachate controls, including an associated wastewater treatment facility. About 60 to 70% of Japan's landfills are of this type.

Table 32 Landfill Sites

	Number of Landfill Sites by Type					Total Area (1,000m²)	Total Capacity (1,000m³)	Capacity Remaining (1,000m³)
	Valley	Sea area	Water area	Flatland	Total			
1976	1,616	34	87	942	2,679	51,427	378,081	244,623
1977	1,614	34	76	914	2,638	43,514	331,899	194,735
1978	1,707	39	59	872	2,677	51,946	392,565	239,191
1979	1,583	39	54	799	2,475	46,625	425,761	214,168
1980	1,600	36	50	796	2,482	52,086	356,109	191,945
1981	1,619	40	48	779	2,486	53,581	403,156	181,578
1982	1,612	36	46	778	2,472	53,929	377,583	175,975
1983	1,638	34	41	766	2,479	55,094	382,728	170,795

Source: JICA

*Fly ash is made into cement bricks immobilizing toxics to
reduce leaching from landfills and facilitate land
reclamation after landfills are full.*

*Landfill sites are used to store sorted waste (including liquor bottles
shown here) to be collected for recycling.*

CHART 9a Final Disposal Sites for Waste (Type A Landfill)

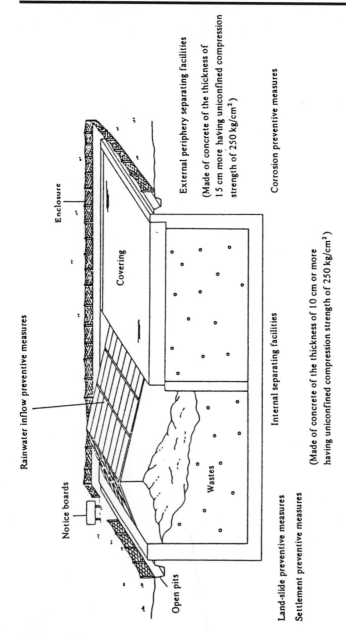

Source: JICA

CHART 9b Final Disposal Sites for Waste (Type B Landfill)

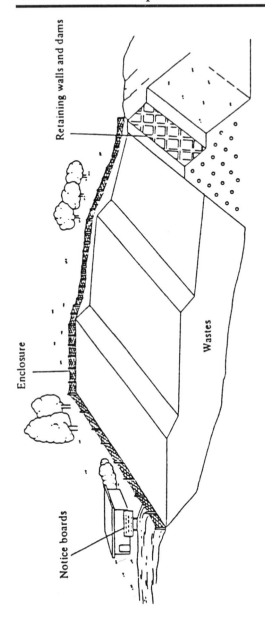

Source: JICA

CHART 9c Final Disposal Sites for Waste (Type C Landfill)

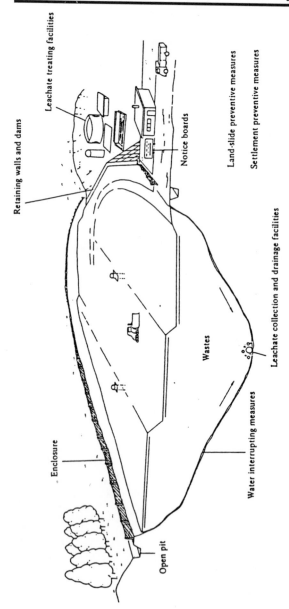

Source: JICA

As one would expect, except for the "sea reclamation" sites, most landfills are located in rural areas. All are publicly owned.

REGULATIONS

Landfills may contaminate soil, underground water and streams by leachate. They must meet structural and technical standards and obtain permits. According to JICA:

It is important that the selection of a landfill site is in accord with an overall land utilization program and that a long-range and extensive administration system is established to guarantee environmental protection after completion of the landfill.

Regulations include enclosing the site, preventing leachate from polluting surface and groundwater, ventilation of landfill gases, "necessary measures to prevent leaking of bad smells out of the disposal site" and vector control.

Both landfills and landfill operators are required to have operating permits. When Mr. K. Onogawa, Deputy Director of Water Quality at the Environment Agency, was asked if all landfills in Japan had permits, he at first seemed surprised at the question and then answered, almost indignantly, "Of course!" (This is in contrast to the U.S., where less than 60% the landfills operate legally.) When capacity is reached, landfills are capped and usually reclaimed for reuse. The Machida landfill will eventually become a recreational site including tennis courts, a baseball field, a soccer field and a track. However, Japanese officials have begun looking for alternative uses because, as Mr. Onogawa says, "We don't need so many golf courses."

Japan's law 137 enacted in 1970 states that "the structure [of a landfill] shall prevent water contained in the waste, wastewater or waste-fluid caused by the waste treatment from leaking or permeating into the underground." New, strict landfill regulations, including liner requirements, went into effect in 1977. Existing landfills were exempted from the new regulations and are allowed to operate until full. Consequently, not many new landfills have been opened since 1977 as it is "difficult to stop using the old ones" according to Mr. Onogawa. Surprisingly, there is no concerted effort to examine for groundwater contamination from these older landfills and not all even have monitoring wells.

However, according to Mr. Onogawa, Japan is not experiencing severe groundwater contamination problems from these landfills. If a problem is suspected (through neighboring well contamination, visible leachate, etc.), environmental offices will respond with tests, remedial actions or even closure. According to Mr. Onogawa, it seems that groundwater pollution is caused mostly by industrial, not municipal waste. Nevertheless, many municipalities have installed liners in the as yet unused sections of their grandfathered landfills to avoid problems which could lead to closure, since finding a new site would be difficult.

Guidelines for siting for new landfills include provisions for soil conditions, hydrogeology, water quality and usage, precipitation, wind direction, traffic patterns, and location of existing wells and trees. Big projects are subject to environmental impact assessments and public comments. The Ministry of Health and Welfare gives courses for landfill operator certification.

APPENDIX 1

Approximate Metric Conversion Factors

To convert from:	To:	Multiply by:
kilograms (kg)	pounds (lb.)	2.2
tonnes (t; metric tons or 1,000 kg)	short tons	1.1
meters (m)	feet (ft)	3.3
meters (m)	yards (yds)	1.1
kilometers (km)	miles (mi)	0.6
sq meters (m²)	sq yd (yd²)	1.2
hectares (ha; 10,000 m²)	acres	2.5
sq kilometers (km²)	sq miles (mi²)	0.4
cubic meters (m³)	cubic yds (yd³)	1.3
Celsius (°C)	Fahrenheit(°F)	9/5 +32
grams/normal cubic meter (g/Nm³)	grains/dry std. cubic foot (gr/dscf)	0.4
kilocalories/kilogram (kcal/kg)	British thermal units/pound (Btu/lb)	1.8

APPENDIX 2

RISACHAN'S PICTURE DIARY ON RECYCLING

CONTENTS

Japan imports 99% of the natural resources necessary for its own use. Let us
look around. Many items used daily, such
as stationery, sport shoes, cans for fruit
juice, bottles, etc. are produced from
imported resources. In particular, bauxite
which is a raw material for aluminum cans
is 100% imported and iron ore which is
the raw material for steel cans is 99.5%
imported. Therefore, if the import of
natural resources stops, most of the goods
cannot be produced in Japan. Every individual therefore must collect used cans
and bottles, and all of us must expand the
movement to recycle resources by our-
selves.

'Recycling' is the activity to collect reusable materials from discarded waste and
effectively use such materials as resources.
This protects the precious nature of the
earth on which we live.

Let us think over, together with a friend
of ours named 'Risachan', the reuse of
used cans and bottles.

❷

How are cans and bottles manufactured?

Aluminum, which is the material to manufacture aluminum cans, is produced from ore called bauxite. Bauxite imported and unloaded is processed into alumina through various steps. Alumina is electrolyzed and processed into aluminum ingots which are then rolled into aluminum sheets. The sheets are then punched, printed and finally become aluminum cans.

The material for steel cans is iron. Iron is manufactured from iron ore. The first iron produced from the iron ore is called 'pig iron', made by properly mixing the ore with coke, limestone, etc. and melting at about 2,000°C. The subsequent processes are the same as those for aluminum cans.

Raw materials for bottles are silica sand, soda ash, burnt lime and cullet (glass scrap). These raw materials are mixed with a coloring agent and auxiliary raw materials, then melted at about 1,600°C. A necessary quantity of the melted glass is taken out, then sent to a bottle forming machine. Bottles from the bottle forming machine are gradually cooled, inspected and finally become the product.

Glass bottles are classified into two types; (1) 'returnable bottles' (for beer, beverage, etc.) which can be repeatedly used, and (2) 'one-way bottles' (for cosmetics, chemicals, etc.), which can be used only once. Returnable bottles are recovered, cleaned and reused. One-way bottles are reused as a raw material for glass making (i.e. cullet).

Can manufacture

Unloading bauxite   Alumina

Open-hearth furnace   Blast furnace

Converter Electric furnace

Bottle manufacture

Raw material mixing facility   Raw material charger

❹

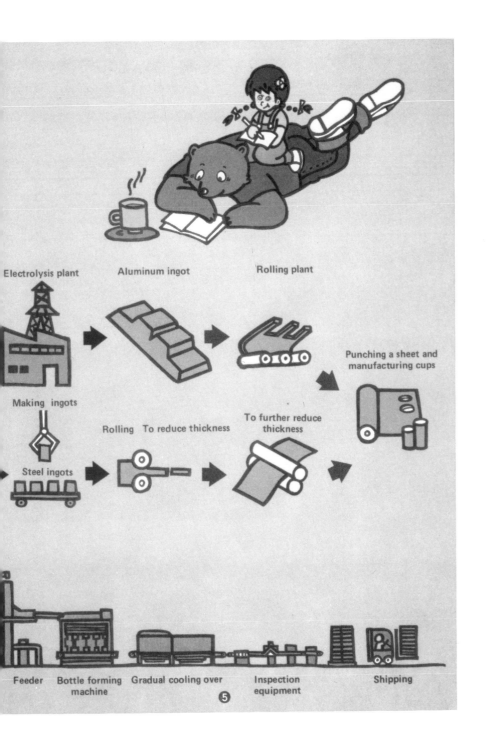

"It doesn't matter. No problem" results in of nature and a loss of resources.

Many used cans have been randomly discarded in various places without re-using. This results in contaminating beautiful nature without effectively re-using valuable resources. The most important point for recycling is to put used cans and bottles into specified waste baskets or bring them back to your home to dispose of through the specified procedure as domestic waste. A little care by individuals will protect the beauty of nature.

a destruction

Many small used cans become a large of resources when they are collected.

Economic activity in Japan has become very dynamic and our standard of living has become high. But, as a result, a large quantity of raw materials have to be consumed. Various products which make our life comfortable and convenient are manufactured by the efforts of many people from such valuable resources.

However, most used cans up to now have been processed at cleaning plants and used for land reclamation. Therefore, valuable resources have not been recovered. From the standpoint of energy-saving, if used cans are recovered and processed for reuse instead of using bauxite and iron ore, about 96% and 65% of the energy can be saved for manufacturing aluminum cans and steel cans respectively.

Nowadays, the active efforts of people to collect used cans and bottles for reuse result in the fact that 26% of used aluminum cans, 39% of used steel cans are recycled, and about 2.6 billion used bottles have been recovered. Let us decide to participate in the movement, then think and talk with each other about all the possible methods of reusing waste.

Recovery

Compressing

Recovery of used cans

Pressing

⑧

amount

Crushing

Melting plant

Aluminum ingots

To the plant

Electric furnace

Product

The earth is our own extensive dwelling

It is important that we have learned how to reuse resources and we can further accomplish this task if we are more careful and exercise a little more effort. Moreover, we have learned that Japan has few natural resources, that products are made by the various efforts of many people, and that the beauty of nature will be destroyed if waste is continuously discarded in a random way. What would you feel like if your house was filled with waste? Day-to-day living would hardly be pleasant. This is the same for the earth. So, in order to protect the precious beauty of nature and the limited amount of natural resources, let us continue our individual efforts to recover discarded used cans and bottles.

Recycling of energy

Use of electric power

Day care home for the aged

Greenhouses

Use of heat

Hot water swimming pool

APPENDIX 3

SOURCES

On-site research for this report took place in Japan between September 9, 1986 and October 15, 1986. This included interviews with more than forty members of Japan's government, waste management industry and academic community. Institutions represented in these interviews were:

Clean Japan Center
Environment Agency
Environmental Science Institute of Hyogo Prefecture
Institute of Public Health
Japan Industrial Pollution Control Association
Kyoto University
Ministry of Health and Welfare
Plastic Waste Management Institute

These interviews were supplemented by the following documents that were provided to INFORM:

Clean Japan Center, *Recycle Life*

Clean Japan Center, *Recycling '86*

Clean Japan Center, *Let's Think About Rubbish*

Clean Japan Center, *Risachan's Picture Diary on Recycling*

Combustion Workshop Proceedings, Machida, Japan, September 12-13, 1986

Dioxin 86 Symposium Proceedings

Environment Agency, Government of Japan, *Illustrated White Paper on the Environment in Japan* 1985

Environment Agency, Government of Japan, *Introduction to the Environment Agency of Japan*

Environment Agency, Government of Japan, *Quality of the Environment in Japan* 1985

Hiraishi T., Chemicals Programme Coordinator, Environment Agency, Japan, "Environmental Administration in Japan and Future Challenges," July, 1986

Japan International Cooperation Agency, Government of Japan, *Solid Waste Management and Night Soil Treatment* (2 vols.) 1984

Okuzawa, S., Nomura Research Institute, "Recent Environmental Issues Related to Hazardous Wastes (Materials) in Japan: A Review," Proceedings of the First U.S. — Japan Workshop on Risk Management, October 28-31, 1984, Tsukuba, Japan

Plastic Waste Management Institute, *Plastic Waste*, Japan, 1985

Three international dioxin conferences were attended, in Machida, Fukuoka and Kyoto. Two landfills were visited (Machida and Mitaka) and eight of the most advanced waste-to-energy plants (built by four different vendors) were studied:

Plant	Manufacturer
Machida City	Ishikawajima-Harima Heavy Industries
Kumamoto City/ Western Plant	Takuma Co.
Sohka City	Takuma Co.
Fukuoka City/ Eastern Plant	Takuma Co.
Mitaka City	Takuma Co.
Ohmiya City	Kawasaki Heavy Industries
Fuji City	Kawasaki Heavy Industries
Yokohama Hokubu	Mitsubishi Heavy Industries/ Martin

About the Authors

Allen J. Hershkowitz, Ph.D.

Project Director, Municipal Solid Waste Research, INFORM.

Dr. Hershkowitz joined INFORM's staff in September 1984, and is the author of INFORM's report, *Garbage Burning: Lessons from Europe.* He was an Assistant Professor of Political Economy at the City University of New York from 1978 to 1984, and now still serves as an adjunct. He is Chair of the New York State Department of Environmental Conservation's Board of Operating Requirements for Municipal Solid Waste Incineration.

Dr. Hershkowitz has researched and lectured on resource recovery in Europe and Japan, and has had articles published on energy and solid waste issues in the *New York Times, Newsday, Nation, City Limits,* and *American Book Review.* He has a Doctor of Philosophy degree in Political Economy with an emphasis in technological and environmental public policy, as well as a Bachelor's degree, from the City University of New York.

Eugene Salerni, Ph.D.

Research Coordinator, New York State Legislative Commission on Solid Waste Management

Dr. Salerni has directed and conducted the Legislative Commission's research activities since its inception three years ago. Among his publications with the Commission is the 1986 report, *Where Will the Garbage Go? New York's Looming Crisis in Disposal Capacity.*

Dr. Salerni received his Doctor of Philosophy degree in Urban and Environmental Studies from Rensselaer Polytechnic Institute.

OTHER RELATED INFORM PUBLICATIONS OF INTEREST

GARBAGE BURNING: LESSONS FROM EUROPE
Consensus and Controversy in Four European States

by Allen Hershkowitz, Ph.D.

This report on the policies and regulations at five garbage-burning plants in Norway, Sweden, West Germany and Switzerland identifies many practices in worker training and in monitoring and controlling air pollutants that are far superior to such practices in the U.S.

(1986, 53pp).....$9.95

TRACING A RIVER'S TOXIC POLLUTION
A Case Study of the Hudson, Phase II

by Steven O. Rohmann, Ph.D. and Nancy Lilienthal

This report extends Phase I of INFORM's study of 26 chemicals polluting the Hudson River. Researchers evaluated over 17,000 analyses of samples of the Hudson River water, fish, and sediment.

(1987, 209pp).....$19.95

set...............$25.00

TRACING A RIVER'S TOXIC POLLUTION
A Case Study of the Hudson, Phase I

by Steven O. Rohmann, Ph.D., with Roger L. Miller,
Elizabeth A. Scott and Warren R. Muir, Ph.D.

This report provides the first inventory of toxic discharges into an entire U.S. waterway. It identifies 185 industrial and sewage treatment plants discharging major toxic chemicals into the Hudson River and its tributaries.

(1985, 150pp).....$12.00

PROMOTING HAZARDOUS WASTE REDUCTION
Six Steps States Can Take

by Dr. Warren R. Muir, Senior Fellow and
Joanna D. Underwood, Executive Director

This report, the result of more than five years of research, details why hazardous waste reduction should be an aggressively pursued management approach. It identifies six initiatives that states can adopt in developing their own waste reduction program.

(1987, 21pp).....$3.50

CUTTING CHEMICAL WASTES: What 29 Organic Chemical Plants Are Doing to Reduce Hazardous Wastes

by David J. Sarokin, Warren R. Muir, Ph.D.,
Catherine G. Miller, Ph.D., and Sebastian R. Sperber

This report offers the first in-depth demonstration of the efforts in a major U.S. industry to reduce the generation of hazardous wastes at source.

(1986, 535pp).....$47.50

FORTHCOMING PUBLICATIONS

FACTS ABOUT U.S. GARBAGE MANAGEMENT
Problems and Practices*

A summary of significant information about the problems and potential strategies in handling municipal solid wastes in the U.S., by Joanna D. Underwood, Executive Director and Allen Hershkowitz, Ph.D.

Full-Length Study of U.S. Garbage-Burning Plants*
by Allen Hershkowitz, Ph.D.

* Working title. Prices and approximate dates of release of publications are available upon request.

6381